AF839243

Iulia Nistor
Intentionality and the Paradox of Fiction

Philosophy

Iulia Nistor, born in 1985, is working in philosophy of mind and epistemology. She received her doctorate in philosophy from Universität Regensburg.

Iulia Nistor

Intentionality and the Paradox of Fiction

A Study in the Philosophy of Mind

[transcript]

Dissertation titled "How We Relate: Intentionality and the Paradox of Fiction", Universität Regensburg, 2022

Bibliographic information published by the Deutsche Nationalbibliothek
The Deutsche Nationalbibliothek lists this publication in the Deutsche Nationalbibliografie; detailed bibliographic data are available online at https://dnb.dnb.de

Hermannstraße 26 | D-33602 Bielefeld | live@transcript-verlag.de

Cover design: Kordula Röckenhaus
Copy editing: Katie van Heest
Printing: Elanders Waiblingen GmbH, Waiblingen
https://doi.org/10.14361/9783839478400
Print-ISBN: 978-3-8376-8238-0 | PDF-ISBN: 978-3-8394-7840-0
ISSN of series: 2702-900X | eISSN of series: 2702-9018

Printed on permanent acid-free text paper.

Contents

Introduction

> We will meet this afternoon. And I won't even talk to you about this that I'm writing...
> *Clarice Lispector, Água Viva*

The way in which Lispector has written *Água Viva* seems to be addressing us directly as if in a letter or a diary she will give us one day: formulas such as 'let me tell you' (Lispector 2012, 3), 'when you come to read me' (4), or 'I see that I've never told you' (5) are used in every other paragraph. Simultaneously, however, we come to realise, through passing references such as 'and you, with your habit of wanting to know why' (5), that it is not us she is addressing but some other person she is familiar with. Throughout the text, it never becomes clear who is the person she is writing to – perhaps they exist, perhaps they are a fictional character, or this it is just a way of writing an inner monologue. Furthermore, we are tempted to identify the voice of this monologue with that of Lispector herself, but the narrator might be fictional as well. The experience of reading *Água Viva* is one in which, despite the insistence on the 'you', we are held at arm's length. We do not empathise with either the person addressed or the person addressing. Concerning the former, we feel rather like eavesdropping on a private conversation, and the intimacy we access is not of the exciting sort but merely unrelated to us. Concerning the latter, it is difficult at times to follow the fragmented lines of this inner monologue, ranging from far-fetched ideas to the trivialities of daily life.[1] For Lispector, what is at stake here seems to be something other than drawing the reader into the story.

To be immersed in fictional scenarios and empathise with characters in books and films is part of our everyday experience. In philosophical discourse, this phenomenon gave rise to questions concerning the relation between

1 Cf. Benjamin Moser in the introduction to Lispector's book (2012, viii).

emotion and fiction and, more recently, was even formulated as a paradox: the so-called paradox of fiction, which originated in a paper by Colin Radford and Michael Weston entitled 'How Can We Be Moved by the Fate of Anna Karenina?' (1975). The modern debate subsumed under the paradox of fiction discusses the seeming inconsistency of our emotional responses to fiction and is typically formulated in three premises stating, in slightly different ways, that (1) we often experience emotional responses to fiction, (2) in order to have an emotion, we need to believe in the existence of the object of our emotion, and (3) we do not believe in the existence of fictional characters or situations. Attempts to solve the paradox have typically sought to show that one of the premises is wrong or can be reformulated in such a way that the three claims do not add up to a contradiction. However, instead of understanding the paradox as a problem to be solved, we can also see it as a tool for examining the different concepts involved. The paradox has already proved useful for refining our concepts of emotion and fiction; it helped, for instance, to refute the narrow cognitivism of emotions, which led to a 'paradigm shift' in emotion theory.[2] Next to the concepts of emotion and fiction, the issue of intentionality lies at the core of the paradox. Our emotional responses to fiction are typically formulated as directed at fictional characters and situations: we feel pity *for* Anna Karenina, and we are afraid *of* Nosferatu. However, the concept of intentionality has not yet been questioned directly in the context of the paradox.

Intentionality should not be confused with 'intention', the determination to act in a certain way, but is a technical term in the philosophy of mind paraphrased as the 'aboutness' or 'directedness' of our mental states. The concept originates in medieval philosophy and was reintroduced to the modern debate by Franz Brentano in 1874, who claimed that it is the mark of the mental, distinguishing mental from physical states. Nowadays, few theorists would still attribute intentionality to all mental states, but it remains a crucial concept for addressing the property by virtue of which our mental states are directed at or are about something. However, although the term 'intentionality' is widely used, and many debates are concerned with it, there seem to be different understandings of the term. This suggests that it needs clarification.

In my study, I will use the paradox of fiction as a perspective tool for examining the concept of intentionality. This will not only help to make implicit understandings of intentionality visible but also offer us a two-fold perspective

2 Cf. Vendrell Ferran (2014).

on the subject: we have to account not only for the intentionality of our emotional responses to fiction, but also consider the intentionality of the artefacts, such as books and films, through which we encounter fiction, since it is usually through a form of presentation that we learn about fictional characters and situations. Words and images are also 'about' something – they point to or are directed at something beyond themselves. The paradox thus prompts us to examine two instances of intentionality: the intentionality of emotions and that of artefacts. The intersection of these two perspectives can help identify relevant aspects of intentionality, make inconsistencies visible, and determine the requirements that a theory must meet. In this sense, I believe that the paradox of fiction can serve as a tool and touchstone to further our understanding of intentionality.

In the first chapter, I present the paradox of fiction as a perspective tool, which can help question existing beliefs and assumptions, as well as determine requirements concerning the concepts involved. First, I will summarise and present the paradox and its most prominent solutions and show that the concept of intentionality was neglected in the context of emotional responses to fiction (Ref). I then propose to analyse the concept of intentionality by using two perspectives that emerge from the paradox: (A)' whether there is a difference between the intentionality of thought and that of emotion, and (B)' whether there is a difference in intentionality depending on whether it is directed at a factual or a fictional state of affairs. These two perspectives will serve as my guideline for considering the understanding, implications, and consequences of intentionality in the context of Ref. For perspective (A)', I am going to discuss the concept of affective intentionality, and for perspective (B)', the concept of derived intentionality. I close the first chapter with several clarifications on how I will use some of the essential terms involved.

In the second chapter, I examine the aspect of intentionality concerning perspective (A)', the difference between the intentionality of a thought and that of an emotion. I first present the argument concerning the reduced behavioural output of Ref in discussing Kendall Walton's (1990) 'three men' scenarios – Henry, Charles, and Aaron. This argument claims that we typically do not act upon the emotions we experience towards fictional characters or situations and is based on what seems toconstitute one of the differences between thoughts and emotions – namely that emotions involve tendencies to act. I analyse Walton's understanding of intentionality and further outline the role intentionality has generally played, and continues to play, in our concept of emotions. I then consider a promising concept regarding the intentionality of emotions: that of affective intentionality, as it is presented in the accounts of Sabine Döring, Bennett Helm, and Jan Slaby. The concept of affective intentionality claims that emotions have a different kind of intentionality – an intentionality sui generis. After discussing the advantages and disadvantages of applying affective intentionality to emotional responses to fiction, I highlight a different general understanding of intentionality in the accounts discussed.

In the third chapter, I address the aspect of intentionality that concerns (B)': the difference between a mental state being directed towards a fictional or a factual state of affairs. I first discuss Radford's (1975) 'tragic sister', an example that is one of the sources of the debate around the paradox. Based on the analysis of this example, along with some considerations of Frank Zipfel (2018), I insist on the importance of the presentational aspect of fiction, namely that we are always confronted with artefacts when dealing with fiction. Whether we read a book, watch a film, or attend a performance, we encounter fictional characters and situations through a form of presentation. Consequently, we must inquire into the intentionality of artefacts if we are to address how mental states are directed to (or through) something that is itself directed. For this purpose, I consider Searle's (1983) concept of derived intentionality. Based on the intuition that words and other signs do not mean something by themselves but are assigned their meaning by their maker or recipient, Searle distinguishes a derived intentionality of language, words, images, and so on, from the original or intrinsic intentionality of the mind. I first present Searle's concept and then discuss its advantages and disadvantages in explaining Ref. The problematic aspects of the concept of derived intentionality, I point out, are its close link to the author's intention, as well as the implications of the term 'derived'.

After examining the concepts of affective and derived intentionality, I turn, in the fourth chapter, to what I refer to as the 'general' understanding of intentionality. There are different ways in which the term 'intentionality' is understood and used, associated with different traditions of thought. Instead of tracing these differences back to these traditions, I propose a different approach: to analyse the different paraphrases used to describe intentionality and to determine from this analysis whether fundamentally different understandings are at play. I examine descriptions of intentionality as content and directedness, as a property of mental states, or as a capacity of the mind. These distinctions provide the general structure of the chapter. After presenting a short history of the concept, I discuss intentionality as content and directedness.

For this purpose, I look at Brentano's original formulation of the term and its subsequent reception. I then discuss intentionality as a property or capacity, drawing on discrepancies found in the course of the preceding chapters, particularly among proponents of affective intentionality. While the descriptions of intentionality as content and directedness may be used synonymously and reflect the twofold nature of intentionality, the descriptions of it as a property of mental states or as an activity of the mind must be differentiated.

In the final chapter, I summarise the insights gained from examining intentionality through perspectives (A)' and (B)', and from analysing the paraphrases used to describe it. Building on these results, I propose a set of requirements that a theory of intentionality should meet. Furthermore, I introduce a new paraphrase for intentionality that emerged during this study: the notion of *relatedness*. 'Relatedness' can refer both to *what* we relate (as in 'narrate'), and that we *relate* (as in 'refer', but also in the expression 'I can relate to that'). Any adequate theory must both account for the various instances of intentionality and preserve the necessary distinctions specified in the requirements.

Before proceeding to the text, some acknowledgements are due. I would like to express my sincere gratitude to my doctoral supervisor, Hans Rott, for his guidance, insight, and steady support throughout this project. I am also grateful to my second reviewer, Tim Kraft, for his careful reading, valuable feedback, and encouragement. I am especially indebted to the Philosophical Colloquium (FoKo) at the University of Regensburg, in which regular discussions provided an essential forum for developing and refining many of the arguments presented here. In particular, I would like to thank Verena Wagner, Eva-Maria Konrad, Stefan Ruhland, Christiana Werner, Holger Leuz, Christian Wirrwitz, Mario Günther, and Simone Aicher for their thoughtful comments and critical engagement. I am likewise grateful for the stimulating conversations on related topics with participants in the Whitney Independent Study Program in New York and at the Philosophical Colloquium at CUNY last year, which helped hone several of the ideas formulated in this work.

1. The paradox of fiction as a perspective tool

> There are no forks, no spoons
> in the dwarfs' house,
> no lunches, no dinners.
> Actually, everyone's had enough
> of the morning glories, the yellow
> evening poppies,
> the apricots turned to brandy,
> – the artificial paradise
> of tiny old men.
> *Nina Cassian, Project for a possible / impossible fairy tale*

The debate around the paradox of fiction has turned out to be more than a matter of solving a puzzle. The paradox was, and still is, fruitful in making visible the inconsistencies in our claims, and in determining requirements for our theories. Most notably, we can observe that during the debate – now spanning roughly 50 years – the concept of emotion has changed. At the time the paradox was first formulated and until the end of the 1990s, cognitive emotion theory was predominant, holding that a belief in the existence of the object of an emotion is constitutive of that emotion. Nowadays, even cognitivist theories of emotion no longer uphold this claim, but have shifted from a narrow to a broad cognitivism, for which a cognitive component, less than a belief (e.g., a thought or imagination), is a sufficient requirement for emotion. Íngrid Vendrell Ferran (2014) even speaks of a paradigm shift, as recent results from emotion theory and neuroscience have led to a widely acknowledged new stance concerning emotion. She claims that this shift was caused and fostered by the debate around the paradox of fiction.[1] With such a revision of the term 'emotion', proposition (2) – that in order to have an emotion we need to believe in the existence of the object our emotion is directed towards – can no longer be upheld, and hence the paradox can be considered solved.[2]

This does not mean that the paradox was a futile problem, as it proved fruitful for examining the underlying notions and related issues. It not only helped to bring about this paradigm shift in the theory of emotion but it has also led to various revisions concerning fiction,[3] doxastic conditions,[4] and the re-evaluation of the role of simulation and imagination in our cognition.[5] Although the paradox can be technically considered solved – in particular, no one accepts proposition (2) any longer – many theorists do not leave matters there. We should still care about the paradox of fiction, claims Robert Stecker (2011), because considering other solutions to the paradox can lead us to better understand the nature, variety, and functional role of emotional responses to fiction. He speaks of the paradox as a 'valuable tool' (308), and Konrad, Petraschka, and Werner agree with him, pointing out that 'there were, and there still are, a number of questions raised in the context of the paradox of fiction [...] which are of equally vital interest for cognitivists and non-cognitivists' (2018, 194).[6] Along the same lines, Vendrell Ferran (2014) calls the paradox a 'heuristic tool', arguing that it offers a concrete perspective on our engagement with fiction and an angle from which to structure different answers to this question.[7]

1 'Die Diskussion um das Paradox hat sich als sehr fruchtbar für die heutige Philosophie erwiesen, eben weil sie den besagten Paradigmenwechsel in der Gefühlsforschung motiviert und beschleunigt hat' (2014, 333–334).

2 There have been different solutions proposed to the paradox, of which I give a summary in 1.1. However, the revision of the concept of emotion and its requirements has been the most widely acclaimed solution and has led to the question of whether we should still care about the paradox of fiction, which has been addressed, for instance, by Stecker (2011) and Moyal-Sharrock (2009).

3 Cf. Moran (1994), Zipfel (2012).

4 Cf.Schaper (1978),Werner (2020).

5 Cf. Walton (1990), Meskin and Weinberg (2003), Nichols (2007).

6 Konrad, Petraschka and Werner are referring to cognitivist/noncognitivist positions in the theory of emotion.

7 'Vielmehr sollte die Diskussion um das Paradoxon als ein heuristisches Mittel verstanden werden, die Frage nach unserer Beschäftigung mit Fiktionen aus einem konkreten Blickwinkel zu behandeln und verschiedene Antworten auf diese Frage zu sortieren' (Vendrell Ferran 2014, 316).

What makes the paradox so fruitful for theoretical debate? Different answers to this question emerge. Emotional responses to fiction are a phenomenon of our daily life, about which we have strong intuitions, and which are compelling to engage with to exemplify or test our theoretical claims. Furthermore, I believe that its potential lies in the different perspectives it affords. In emotional responses to fiction ('Ref' for short), subjects such as emotion, fiction, belief, intentionality, and behaviour are thoroughly intertwined. This entanglement allows us to approach the same phenomenon from different angles – or rather to examine one subject through the lens of another. We can, for instance, analyse emotion from the perspective of belief by considering the challenge posed by fictional characters as objects of our emotions. Alternatively, we can examine fiction from the standpoint of behaviour: although Ref involve empathy for fictional characters, this does not typically result in corresponding real-world action.

As we are no longer dealing with a paradox proper, the propositions traditionally forming the paradox are often only mentioned out of historical interest. The ongoing debate, which originated as the paradox of fiction, is now more commonly referred to in terms of the phenomenon itself, namely, 'emotional responses to fiction' or 'fictional emotions'.[8] Instead of a set of propositions, other formulations have been proposed that are more apt to the respective focus or question under discussion. Meskin and Weinberg (2003), for instance, do not believe that solving the paradox by rejecting one of its premises suffices, and they propose a list of explananda that every theory trying to solve the paradox must comply with.[9] I want to propose yet another way of reiterating the challenges posed by the paradox: by taking it as a perspective tool, I will formulate two main perspectives that Ref offer. That is, if we take the relation e (emotion) – f (fiction), we can choose either emotion or fiction as our point of view. In this way, we come to the following picture:

emotion/thought > f

e < fiction/fact

8 E.g., Konrad, Petraschka and Werner (2018).

9 Meskin and Weinberg's explananda concerning Ref are (a) phenomenological and physiological robustness, (b) fiction-directed affective intentionality, (c) behavioural circumscription, and (d) intimate yet ambiguous relationship to nonfictive affect and emotion (2003, 20).

These two perspectives are presenting us with two questions, respectively. From the point of view of emotion, we can ask whether there is a difference if a thought or an emotion is directed at a fictional state of affairs. From the point of view of fiction, we can ask whether there is a difference when an emotion is directed at a factual or fictional state of affairs. Both perspectives are central to Ref and offer insight into the interrelated subjects from different angles.

(A) Is there a difference between a thought and an emotion directed at a fictional state of affairs?
(B) Is there a difference whether a mental state is directed at a factual or a fictional state of affairs?

I want to use these two perspectives to look at a subject that has not yet been the focus of the debate but, I think, lies at the core of the paradox: that of intentionality. Intentionality is described as the capacity of our mind or the property of our mental states to be directed to or about something.[10] In considering the paradox of fiction, it is important to address not only the concepts of emotion and fiction, but also that of intentionality. Intentionality describes the relation between the two: that our emotional response is *about* a fictional state of affairs. The paradox thus provides a useful framework for examining the nature of intentionality. Formulating these perspectives and using them as a tool to analyse intentionality has at least two merits. On the one hand, these two crucial questions help us navigate the vast domain of intentionality. I treat them non-exhaustively; that is, I focus on two well-known examples from the discourse around the paradox and discuss two promising theories of intentionality accordingly. On the other hand, although debates and theories from the two domains – intentionality and the paradox of fiction – can be addressed only selectively in the present study, the conjunction of the two perspectives promises to offer valuable insights into the nature of the problem, and help us formulate requirements a theory of intentionality has to meet.

Prior to addressing these two perspectives in chapters 2 and 3, I introduce the paradox of fiction and its most prominent solutions. I then elaborate on why intentionality, as stated earlier, lies at the core of the paradox, and show that it has nevertheless been neglected in this context. Finally, I apply the two

10 Various uses of the term will be summarised at the end of this chapter, while different paraphrases that point to potentially diverging general understandings of intentionality will be discussed in detail in chapter 4.

perspectives (A) and (B) more specifically to intentionality and clarify how certain terms are used in this study.

1.1 The paradox of fiction

One of the main reasons why we value works of art such as plays, films, and novels, is the emotional engagement they prompt in us. We do not read *Anna Karenina* for the same purpose as we read a newspaper – to gather information – but rather to share the character's destiny, to be 'afraid for her' or 'happy for her'. We take pleasure in the emotions we experience when reading books or watching films. This is part of our everyday experience, but turns out to be problematic with regard to philosophical explanation. The difficulties arise from the fact that we are moved by something we know to be fictional, namely, the situations and characters in fictional narratives. Formulating emotions as directed towards their objects seems to imply that we believe in the existence of those objects. At least this was the original stance of cognitive emotion theory, which was widely accepted until the 1990s. To preresuppose, alternatively, that we believe the stories and characters of a book or film to be imaginary would be an argument for the irrationality of our emotional responses to fiction. Colin Radford initiated the debate around the paradox of fiction or the paradox of Ref with his essay 'How can we be moved by the fate of Anna Karenina?' (1975), claiming that our responses to narrative fiction are inconsistent and irrational. The paradox is typically formulated in three premises:[11]

(1) We have genuine emotions towards fictional entities.
(2) To have a genuine emotion towards an entity, we must believe that this entity exists.
(3) We do not believe that fictional entities exist.

The formulation as a paradox holds only if we take into account a cognitivist position concerning emotion, claiming that emotions are based on, or entail in some way, a judgement or belief: cognitivists believed, for example, that fear entails a belief or judgement that one is in danger. If confronted with a snake,

11 I present the recent formulation of the paradox by Konrad, Petraschka and Werner (2018). It has been formulated in slightly different ways, for example, by Carroll (1990), Levinson (1997), Yanal (1999), Lamarque (1996), and Vendrell Ferran (2014).

one typically has beliefs about the snake (e.g., that it is in front of us, that it is venomous); one can have a range of feelings from discomfort to alarm; one might have an increased heart rate and tensed muscles, and one might tend to act or behave in certain ways. The extent to which one is inclined to act implies an evaluation of whether or not one is in actual danger. These aspects are part of our emotional experience, and one of the debated questions in theories of emotion regards which of these components are constitutive. The difference between cognitivist and non-cognitivist positions becomes evident when we consider Ref. Seeing Indiana Jones falling into a cave full of snakes or reading about Hercules fighting the Hydra, growing three new heads every time one is cut off, we can be horrified, even have physiological reactions. However, we typically would not take any specificactions directed at the object of our emotion. Cognitivists would say that, in this case, the belief in the existence of the object of emotion is lacking and, thus, we do not evaluate the situation as dangerous to us. This is why earlier positions in the debate around the paradox of fiction typically reject (1), arguing that our Ref are not genuine emotions, that they are 'quasi-emotions' or, in some way, lesser than emotions directed towards factual situations. Levinson formulates the philosophical question concerning Ref as follows: 'How can we coherently have emotions for fictional persons or situations, given that we do not believe in their existence?' (2006, 38). Nowadays, the tendency is rather to reject (2), questioning whether the belief in the existence of the object of emotion is a necessary component of emotion by claiming that a thought or imagination can be sufficient for having a genuine emotional response.

Solutions to the paradox primarily consist of rejecting or refining one or more of the formulated propositions (1) – (3). To give an impression of the complexity of the debate and how different subjects are intricately entangled in it, I briefly present some solutions to the paradox, ordered according to the propositions they criticise or refine.[12]

There are two general possibilities for rejecting proposition (1): the first one questions the status of the emotions experienced towards fiction. This can be accomplished by either claiming that Ref are not 'real' emotions or in some way less substantial. A more sophisticated theory, building on the idea that Ref are not ordinary emotions, claims that they are make-believe emotions or quasi-emotions (Walton, 1978, 1990). Such emotions are generated only in special

12 This overview is based on Lamarque (1996), Levinson (1997), Neill (2009), Vendrell Ferran (2014), and Konrad, Petraschka and Werner (2018).

contexts, namely in circumstances in which we are engaging in games of make-believe. Hence this theory having been called the *make-believe approach*, or the *imaginary solution*. According to Walton, in experiencing make-believe emotions, we do not feel them to be illusory: to us, they seem real and substantial. The second possibility of rejecting (1) is to query whether the object of our Ref is fictional. The object of our emotion is, according to this so-called *surrogate-object solution*, not the fictional person or event, but the work – its performance or description. The object of the emotion could also be a 'shadow-object', such as real people from our experience resembling the represented character. According to the surrogate-object solution, our emotions are not directed to fictional objects but to 'real' ones, both existent and believed to be existent, which act as a surrogate (e.g., Lamarque, 1981).

The rejection of proposition (2) is mostly concerned with arguments from the theory of emotion. The claim in question is that emotions are based on, or entail, judgement or belief, formulating a doxastic condition in which we presuppose the existence of the objects our emotions are directed towards. This can be solved by arguing that a cognitive engagement, less than a judgement or belief, suffices to generate an emotion, or that some types of emotions do not require any cognitive or existential commitments at all. The first proposition is based on the *idea* or *thought approach* (Carroll, 1990;Lamarque, 1996; Yanal, 1999; Meskin & Weinberg, 2003), which asserts that mere thought or description is sufficient to generate emotion. We can be frightened by a description of Nosferatu despite not believing in his existence, because 'vivid imagination' becomes a substitute for belief.The second claim that some types of emotions do not require a cognitive component at all is formulated by the *startle approach* (Choi, 2003). This theory classifies Ref as mere physiological responses: a noise or a movement in the dark can elicit a shock or startle us. Other accounts have focussed on the doxastic requirements of emotion (Schaper, 1978; Werner, 2020).

Although proposition (3) has been widely accepted in recent debates, we should also consider objections to this view. The better-known claim can be traced back to Samuel T. Coleridge's term 'suspension of disbelief'.[13] It suggests that spectators temporarily put aside their rational disbelief in order to

13 The term is mentioned in his book *Biographia Literaria* or, *Biographical Sketches of my literary Life and Opinions* (1917) and is often referenced by film theorists influenced by Christian Metz (1972).

allow themselves to believe in the existence of fictional objects. Another possible rejection of proposition (3) is the *surrogate-belief solution*, which claims that, for Ref, it suffices to believe that characters and circumstances exist in a fiction world (Neill, 1993; Yanal, 1994). From a different angle, Suits (2006) claims that we actually believe in fiction and only act differently because of the context in which we experience it, such as a cinema hall.

Another categorisation concerning the different solutions of the paradox has been made by Vendrell Ferran: she distinguishes between doxastic and practical rationality[14] to show how the questions around the paradox shifted after the revision of the concept of emotion, namely the shift to broad cognitivism. She argues that the foremost effect of this shift was to resolve the doxastic contradiction of the paradox – that we can believe and at the same time not believe proposition p – and she structures her paper around the solutions that were reached before and after. Before the shift, the questions concerned doxastic rationality, whereas now they instead revolve around practical rationality, that is why Ref do not tend to motivate actions (Vendrell Ferran 2014, 315). New accounts concerning Ref analyse, for example, desires about fictional states of affairs (Düringer, 2018) or consider to what extent video games challenge the premise that we cannot interact with fiction (Van de Mosselaer, 2018).[15] However, the debate is not narrowing; on the contrary, it is expanding. Since the discussion is no longer restricted to the purpose of solving the paradox, the latter is taken rather as a point of departure.

I am going to use the paradox of fiction as a perspective tool to look at the concept of intentionality. Until now, to reject proposition (2) was tantamount to questioning the concept of emotion or the doxastic conditions implied. However, there are also implications for intentionality at play, namely that we understand emotion as being directed *towards* something. We formulate emotions as directed towards an object of emotion, and this directedness – whether or not the object exists – must itself be examined. Reading through various accounts concerning the paradox, there appears to be an implicit understanding of intentionality that is neither explicitly addressed, nor questioned. The intentionality of our Ref has not yet been adequately taken into account. I sug-

14 She distinguishes between doxastic and practical rationality for analytical purposes, noting that they are intertwined. Vendrell Ferran (2014, 315n6).

15 Cf. the coedited volume by Konrad, Petraschka and Werner (2018) for a recent collection of new papers around the paradox of fiction. They also supply us in their introduction with a comprehensive bibliography on the debate from 1975 until 2018.

gest that, just as the concept of emotion was called into question, we must also critically reexamine the concept of intentionality.

1.2 Intentionality neglected

Intentionality is a technical term and is not to be understood in the sense of intention; it has been described as the directedness or aboutness of mental states: the capacity of our mind to direct itself towards things, to be about or to represent them. I will discuss how the term has been paraphrased, and the general understandings of it, at length in chapter 4, so I will not go into too much detail now. Here my purpose is limited to showing why the concept is relevant to the paradox of fiction, and how intentionality has so far been neglected in the context of the paradox. In the following chapter, I will proceed to show how Ref can be used as a perspective tool to consider intentionality.

Why is intentionality relevant to the paradox of fiction? Having already stated that I take intentionality to be at the core of the paradox, I will now elaborate on that contention. Intentionality is directly addressed in proposition (2), but it also influences how propositions (1) and (3) are formulated. Let us look at the three propositions with an emphasis on intentionality:

(1) We have genuine emotions *towards* fictional entities.
(2) To have a genuine emotion *towards* an entity, we must believe that this entity exists.
(3) We do not believe *that* fictional entities exist.

Proposition (2) was mostly read as a theoretical claim about emotion, but it can equally be understood as a claim about intentionality. If we do not focus on 'emotion' but on 'towards', we can read the sentence as a proposition concerned with the intentionality of our emotions and their objects. Emotions are experienced as being directed towards something, and we also formulate them in such a way, which suggests that emotions are accompanied by a propositional attitude. In the case of Ref, this propositional attitude is directed towards a fictional object: we are sorry *for* Anna Karenina and afraid *of* Nosferatu. This raises questions about the ontological status of such fictional objects, and about intentionality itself. I am going to focus on the latter.

Intentionality is discussed in the context of emotions – and in the discourse around Ref – as a means of ensuring their rationality. It is through intentionality that emotions become rationally accessible, that their causes and reasons are disclosed, and their appropriateness and coherence become comprehensible. I am going to elaborate on this point more fully in chapter 2 when I consider the role of intentionality in the theory of emotion. As regarding the paradox, intentionality has been taken as a requirement for it to apply: only if a mental state is directed towards something, and this 'thing' does not exist, can a doxastic contradiction potentially be constituted. Intentionality lies at the core of the paradox insofar as it is partly constitutive of it. That we no longer assume that the paradox strictly applies does not diminish the importance of intentionality in the debate surrounding it, because intentionality is essential to the phenomenon in question – namely emotional responses to fiction. Intentionality is relevant to Ref in two aspects: we must account both for the directedness or aboutness of our emotions and for that of the fictional work that elicits them. Ref make an exceptionally interesting case for discussing intentionality as they lie at the intersection of these two instances of it.

Intentionality has been neglected in the debates around the paradox. Although the concept is sometimes invoked in this context, no line of argument specifically focussed on the concept of intentionality has been conducted. The concept has mostly been used to either ensure the rationality of our emotions or as a requirement to constitute the paradox. Furthermore, when it has been considered, the particular understanding of intentionality has largely remained implicit. Meskin and Weinberg (2003) are among the few who give it a more prominent role by naming fiction-directed intentionality as one of the explananda each theory has to account for: they specify that Ref are experienced as directed at fictional characters, events, and situations, that we describe them in intentional terms (e.g., we are sad *for* Anna Karenina), and that we believe such descriptions to be correct when they are used by others. Aside from being designated as one of their explananda, intentionality is further mentioned at two points in their argumentation. First, while making the simulation approach suitable for Ref,[16] Meskin and Weinberg emphasise that fiction-directed simulated beliefs can accomplish the same functional role as simulated beliefs directed to things that exist. Later, they state that the ability to activate our affect system is not restricted to beliefs but applies

16 For making the simulation approach suitable for Ref, Meskin and Weinberg state, 'We can easily broaden the theory to allow simulated beliefs to have their input not just from what we take to be some target person's beliefs, but also from the contents of fictional works' (2003, 24).

also to PWB representations.[17] In none of the cases is further explanation or analysis of fiction-directedness conducted. Although intentionality is given a more prominent role, and identified as an explanandum, it is not further analysed or problematised as a concept.

There are several ways of dealing with the intentionality of Ref: we could claim that Ref do not need an intentional object, we could argue that Ref provide a different kind of intentionality, or we could question the 'general understanding' of intentionality implied in this context.

To claim that Ref do not need an intentional object is possible, but can only be upheld when we restrict the vast field of Ref to 'reflexive affective responses', such as fits and startles (cf. Choi, 2003). Moreover, the case of moods can be instanced, in which an emotion does not have a specific object it is directed towards. One can object to such arguments by holding that these are exceptional cases and that we also have to account for emotions that are directed to distinct fictional objects; that is, we have to be able to explain the directedness of our empathy towards Anna Karenina and the aboutness of our fear of Nosferatu. One could even argue that cases without intentional objects do not strictly fall under the paradox. However, we are no longer strictly concerned with the paradox, and startles and moods are, although exceptional cases, interesting from the perspective of a general theory of intentionality – for example, in raising the question of how we can understand objectless emotions, and whether there is some kind of (bodily) evaluation involved in reflexive affective responses. It remains vital, however, to be able to account for cognitively complex emotions, such as empathy, admiration, or envy, for which a propositional attitude accompanying the emotion would seem to be necessary, in order to provide the emotional state with its 'about'. Most emotions require intentionality, and Ref are no exception. Although reflexive responses and moods are interesting examples that can help reveal certain aspects they do not allow us to circumvent the challenge of accounting for the intentionality of Ref.

---

17 PWB stands for 'possible world box' and is part of their boxological architecture of the mind. They also place fictional contents in the PWB, explaining that 'it is remarkably easy to make us pity people that we have no personal relationship with whatsoever – one thinks of advertisements for charities directed at third-world hunger, for example – and if we do come to care for such people […], it is only because such advertisements first lead us to pity them. It is no more mysterious (though no less an interesting psychological question!) that we should feel pity' (Meskin and Weinberg 2003, 32).

Another way of dealing with the intentionality of Ref is to argue that the intentionality of emotions is different from that of beliefs and desires. There have been attempts to show how intentionality works differently in emotions, considering emotions as a type of intentional state sui generis.[18] The main difference between affective intentionality and other forms is the unification view of the content and quality of emotions it proposes; that is, the view that the intentionality and phenomenality of emotions are no longer to be seen as distinct, but rather as essentially united in emotional experience.

> In an emotion, an aspect of the world is experienced as significant in a specific respect: In fear, something is experienced as a threat; in anger, something is experienced as an offence; in shame, something about oneself is experienced as a defect (according to social or moral standards); in grief, something is experienced as an irrevocable loss; and so on. (Slaby, 2008a, 431)

This new account of intentionality is a promising means of accounting for the intentionality of emotions, and has not yet been applied to Ref. Consequently, its suitability should be tested.

Finally, we could question the 'general understanding' of intentionality implied in this context. As *intentionality* is a technical term and has, after its introduction by Brentano (1874), been developed further in various traditions of thought, it is not surprising that it may be used in different ways. Moreover, the term has invited interpretation from the very beginning, as Brentano himself described it using different paraphrases. We can observe that in recent years several 'kinds' or 'types' of intentionality have been proposed and begun to form their proper discourses – such as affective intentionality, non-propositional intentionality, and phenomenal intentionality. Indeed, there has been considerable movement in the domain of intentionality recently, and I believe this is due to something similar as that which occurred with the concept of emotion a few decades ago: the existing conception of intentionality proved to be too narrow to account for certain phenomena. To examine the general understanding of intentionality would mean, first of all, paying attention to the respective uses of intentionality and making the relevant implications explicit.

I have shown that the first and second proposition of the paradox directly address the fiction-directed intentionality of Ref. However, our understand-

18 Such are, for instance, Stocker (1983), Goldie (2000), Helm (2000), Ratcliffe (2005), Döring (2007), Slaby (2008a), Montague (2009).

ing of intentionality is also implicit in our description of Ref, that is, in expressions such as having empathy *with* or fear *of* some fictional entity. The formulation per se is unproblematic, as this is due to the predicative structure of sentences. Still, expressed in this way, it may convey the impression that we are 'dealing directly' with fictional entities in our thoughts and feelings. Furthermore, when dealing with fictional entities, we are typically engaging with representations that are themselves directed. Whether in the form of descriptions in a book, images in a film, or representation by other means, it is by virtue of such representations pointing beyond themselves that we come to know something about, for instance, Anna Karenina, and only then can we potentially empathise with her. I maintained earlier that intentionality is relevant for Ref in two respects: we must account, on the one hand, for the intentionality of such emotions and, on the other, for the intentionality of the fictional work that elicits them. The present point concerns the latter, namely, that in Ref we are dealing with representations. This was already pointed out, to some extent, by Weston (1975) in direct response to Radford's initial paper, and later, for example, considered by Moran (1994) and Zipfel (2018). However, our encountering fiction through representations has not yet been discussed with an emphasis on intentionality. There seems to be a similar sense of directedness, or aboutness, in artefacts as there is in our mental states. Can what is usually described as reference or representation in artefacts also be seen as an instance of intentionality?

Although intentionality is typically attributed only to mental states, I want to discuss an account that attributes it to artefacts as well: namely, that of derived intentionality. The concept of derived intentionality was introduced by John Searle (1979, 1980, 1983) as a theory of speech acts and was mostly discussed in the context of artificial intelligence and the computational theory of mind. He suggests that there is a primary/intrinsic intentionality of the mind and a secondary/derived intentionality of language, words, images, and so on. According to this view, artefacts such as films, books, images, and sentences do not mean something by themselves, but through the intentionality of the mind. The concept of derived intentionality proposes how the relationship between the intentionality of mind and that of speech acts (and, by extension, other cultural artefacts) is constituted. I want to apply the concept of derived intentionality to the context of our Ref to see whether it can account for the aboutness of the works that elicit them.

Recognising how many aspects of the paradox of fiction are linked to intentionality, the necessity of addressing them becomes apparent. The most promising accounts concerning the paradox of fiction reject proposition (2). I suggest there are at least three reasons for this: first, it is the only theoretical claim, while the other two are based on the everyday experience of our emotional reactions, our behaviour, and our general beliefs. Second, the cognitivist concept of emotion was too narrow and had to be revised. Third, as I have tried to show, it touches upon questions of intentionality, which have not yet been addressed in this context. Moreover, there seem to be implicit understandings of intentionality at play in the discourse around Ref. In this study I propose using Ref as a tool to look at and refine our understanding of intentionality. I will approach intentionality from the two perspectives (A) and (B), which I formulated earlier, to bring out any implicit understandings of the term and derive a set of requirements that an account of intentionality must satisfy.

1.3 Two 'towards'

To examine intentionality through the lens of Ref, I will now apply the two perspectives previously identified to the aspect of intentionality. The two perspectives that emotional responses to fiction offer gave rise to two questions:(A) whether there is a difference between thoughts and emotions directed at a fictional state of affairs, and (B) whether it makes a difference if a mental state is directed at a factual or a fictional state of affairs. To apply these two questions to intentionality, we must focus on the 'towards' in both cases:

(A)' Is there a difference between the intentionality of thought and that of emotion?

(B)' Is there a difference in intentionality depending on whether it is directed at a factual or a fictional state of affairs?

Concerning (A)', I will first analyse Walton's example of Henry-Charles-Aaron (1990), paying special attention to how he understands intentionality. Then I will discuss affective intentionality, and apply it to Ref. This will be addressed in Chapter 2To answer (B)', first, I propose to examine Radford's example of the 'tragic sister' and then discuss the concept of derived intentionality, with regard to Ref. The decision to discuss derived intentionality stems from the observation that when we are dealing with fiction, we do so through representations. Thus, in chapter 3, I will address the intentionality of works of fiction and other artefacts.

Thus, I will use the two perspectives (A)' and (B)' to examine the intentionality of both emotions and artefacts. Looking at intentionality from these two angles, I believe, will not only provide orientation within this vast field, but will also conjointly offer us a more comprehensive view of the problems involved, ultimately highlighting inconsistencies in our understanding of intentionality and helping to determine requirements and possible directions for revising the concept. Throughout my research, I will pay close attention to the implicit understandings of intentionality in the authors referenced, and I analyse how the notion is employed in various paraphrases in chapter 4.

1.4 Use of terms

To approach intentionality through the lens of the paradox of fiction, the use of several notions requires clarification.

The contemporary notion of emotion underwent several shifts, and it is still under construction.[19] How the term 'emotion' itself should be defined has been discussed since the end of the 19th century,[20] with different theories beingproposed and refined in a dialogue between psychology, neuroscience, and philosophy. For a long time, the opposition between feeling-based and thought-based approaches was predominant, placing either sensation or cognition at the core of their understanding of emotion. Emotion, however, proved to be irreducible to either of these two aspects. One theory has difficulty integrating intentionality, and the other struggles to incorporate phenomenality into its concept of emotion. Nevertheless, Döring and Reisenzein conclude that, at present, 'under the pressure of theoretical arguments and empirical evidence, they have moved increasingly closer to each other' (2009, 204). Most accounts of emotions try to do justice to both of these central aspects. Not only are emo-

19 The prominent shifts in the understanding of emotion occurred during the cognitivist turn in the mid-20th century (cf. Döring 2009b, 13) when analytic philosophy came to analyse emotions in propositional terms and incorporate them in the existing theory of the mind (and, thus, take them seriously), and in a transition from narrow to broad cognitivism that took place at the beginning of the 21st century – as pointed out by Vendrell Ferran (2014) – in which a cognitive component, less than a belief or judgment, came to be considered sufficient for emotion.

20 Often William James's article 'What Is an Emotion?' (1884) is referenced as a starting point of the contemporary debate.

tions not definable according to a single aspect, but it also becomes clear that what lies at 'the core of an emotion' is not the only thing that matters:

> The crucial problem is probably intrinsic to the subject: whatever one might designate as emotion is a multicomponential phenomenon [...] in which the relationships between the component phenomena and the role of each component as a constituent (necessary or sufficient) criterion are far from clear[21]

An emotion is a mental phenomenon associated with phenomenology and intentionality – bodily changes, affectivity, action tendencies, evaluation, and so on. The importance of these different features and their interrelation with each other are still intensely discussed. It remains unclear whether, given the range of emotional phenomena, they can be subsumed under a single theory.[22]

A common distinction in this context is that between emotion and feeling: feelings are located closer to the body, emotions closer to the mind. Feelings include not only bodily sensations, such as headaches, but also bodily changes that are part of more complex emotions, such as the uneasy sensation in our guts when we are afraid of flying. Emotions, however, are often understood as differing from feelings in that they are directed towards something in the world that they represent in a certain way. They are typically considered more complex because they involve more cognitive implications (e.g., in the case of envy or schadenfreude). Emotions and feelings are both mental phenomena that share a phenomenal character, a what-it-is-like to undergo experiencing them; however, emotions – according to this distinction – further represent the object of emotion in its significance for the subject: for instance, a snake as being dangerous. It is often by means of intentionality that emotions are differentiated from feelings. Proponents of this distinction usually understand intentionality as the 'having of an intentional object'; for example, we experience our emotions as directed towards something and typically also formulate them in such a way. In this vein, moods are often subsumed under feelings because they do not have an object they are 'about'.

This distinction between emotion and feeling hinges on the understanding of intentionality implied. In the context of cognitive emotion theory, intentionality is often understood as a propositional attitude. Others who un-

---

21 Frijda and Scherer (2009, 142). They name such components as appraisal, action readiness, tuning of the somatovisceral and motor systems, feeling and control precedence.

22 Cf. Döring (2009b, 12).

derstand intentionality as a form of directedness (e.g., Goldie, with his concept of 'feeling towards')[23] consequently do not draw such a clear distinction between emotion and feeling. The terms would still describe the two ends of the phenomenal spectrum, but the transition seems to be more of a gradual one, in their understanding. For them, intentionality is not a distinction criterion but an aspect that can be attributed to feelings as well: a headacheinvolves awareness of a part of our body, and the uneasy sensation in our guts we attribute to having to enter an aeroplane. In this sense, our feelings are directed to our body or to something in the world. Often the distinction between the cause and object of emotion is highlighted.[24] Another category often deployed is that of (reflexive) affective responses, which is on the bodily side of the scale, but should be distinguished from feelingas it excludes cognitive involvement. Choi uses the term, for example, to show that some of our responses to fiction do not possess intentionality and thereby 'get around' the paradox. At present, the term 'affect' – for which intentionality is not a distinction criterion (it may imply it or not) – is often used specifically to emphasise the bodily nature of emotions.

Unless referring to a particular theory, I will use the term 'emotion' or 'emotional response'. The latter should not be understood as reducing the phenomena to reflexive affective responses, as Choi does, but – on the contrary – as broadening the spectrum of what is generally understood as emotion. As I am going to analyse and question the concept of intentionality, I will not strictly differentiate the terms 'emotion' and 'feeling' or at least not according to the criterion of their being intentional. I will rather follow the intuition that they are two ends of a phenomenal spectrum, the broad term covering the whole of this spectrum being 'emotional response'. The term 'emotion', unless discussed as part of a particular theory, I will use in the same broad sense: synonymously and as a short form for 'emotional response'.

Furthermore, I want to differentiate Ref from emotional responses to art. These descriptions are sometimes used interchangeably in the debate given the fact that most works of fiction are considered works of art. However, it is necessary to separate the debate around the paradox from the debate over aes-

---

23 Cf. Goldie (2002).

24 This was also referred to as the distinction between the particular and formal object of emotion, and goes back to Kenny (1963).Confusing the two sometimes led to claims such as that if an emotion does not have an object, it is not sufficiently motivated (cf. my later discussion of Weston).

thetics, primarily so as not to get entangled with questions around the definition and the appreciation of art. The debate over aesthetics is a normative one and concerned with defining and establishing values: for example, whether we consider a work to be 'artistic', its interpretation appropriate, or what value it constitutes for us. The debate around the paradox, however, is concerned with the nonexistence of fictional worlds and – in my focus – with fiction-directed intentionality, that is, how our emotions can be directed to something non-existent. Second, it is necessary to distinguish between the respective debates because their terms are not coextensive: it might be difficult to think of a work of fiction which is not a work of art; however, in contrast, a lot of works of art cannot be considered works of fiction. A sculpture, painting, or work of video art may be purely abstract, devoid of any narrative that would justify us to speak of it as fictional. Furthermore, the shape, colour, and materiality of a work must often be understood in a concrete way and not as pointing beyond itself to something else. Moreover, a lot of theorists and artists would deny that art's purpose is to affect, or even that art necessarily has anything to do with affect. I will elaborate on the distinction between Ref and emotional responses to art (Rea) further in chapter 3 when discussing the intentionality of artefacts. I will further point out that the discussion of formal aspects, which has been led extensively (although to different ends) in the domain of aesthetics, has been largely neglected in the debates surrounding the paradox,, and argue that this should be reconsidered.

Another term that needs clarification is 'fiction'. Etymologically, it can be traced back to the Latin noun *fictio* (coining, shaping, fashioning, pretence, feigning, etc.) and the verb *fingere* (to compose, create, invent, devise, contrive, imagine, mould, form, shape, produce, etc.). There are two senses of the word. On the one hand, it can be a term that signfies nonexistence, and is associated with falsehood, as when someone says 'His doctoral degree is a fiction', claiming that he has no such degree. On the other hand, it can designate a class of works: works of fiction, which can be novels, stories, dramas, poems, films, songs, or even paintings. Those works exist, of course, but bear a certain element of unreality.[25] Concerning Ref, we refer to the latter sense, in which fiction denotes 'the element of representation, in any art form, that involves what is invented, made up, or imaginary' (Lamarque 2005, 377). The term can apply to objects, as well as to certain kinds of description.[26]

---

25 Cf. Stecker (2010, 163f).

26 Cf. Lamarque and Olsen (1994).

How is fiction ultimately to be distinguished from nonfiction? There is no specific medium for fiction, it can be adopted by any art form. Although there are narrative structures characteristic of fiction, and respectively of nonfiction, this is no criterion of differentiation, as these structures are often intertwined.[27] In literature, biographies, literary journalism, or travel writing can be factually accurate, even as they are shaped in the style of a novel or even a poem. Such texts belong to the genre of creative nonfiction. There are, however, also fictional works that adopt nonfiction techniques and modes, such as journalistic, diaristic, or academic styles. As for the representation of actuality, we cannot take truth or falsity as criteria for differentiation between fiction and nonfiction. We can find sentences or entire paragraphs in works of fiction that can be taken as true assertions, while invented or even falsedetails are to be found in works of nonfiction (e.g., early scientific texts or stories that have subsequently been discredited). Many theories that seek to distinguish between fiction and nonfiction, therefore, focus on how each respectively functions or how each is intended to function. Searle, for example, claims that in contrast to a paradigmatic speech act – that is, one that asserts that something is the case – fiction is the product of *pretending* to assert that something is the case.[28] For Walton, ont he other hand, fictionality is independent of the intentions of the narrator; rather, it is the function of a book, film, and so on, insofar as it is socially recognised as serving as a prop[29] in games of make-believe. Whereas Searle has difficulties with pretended assertions that do not result in fiction (e.g., acts of mimicry, or lies), Walton has difficulty preserving the distinction between a narrative being fictional and its being treated as fictional. Alternative accounts of speech act theorists[30] suggest that fictionality derives from an author's intention that the audience believe the narrated events.

I am going to follow Robert Stecker, who claims that a work of fiction has the proper function of authorising us to imagine things 'for the sake of being imagined without being placed for the purpose of expressing a truth or referring to an item in the actual world' (2010, 170).The adjective 'fictional' refers to a work of fiction or to the fictional status of elements of such works, which include fictional characters and situations. To denote the fictional events in the

---

27 Cf. Davies (2009).

28 Cf. Searle (1975).

29 Props are requisites in plays that guide and sustain imaginings, such as dolls in children's plays.

30 Cf. Lamarque (1996); Currie (1991).

story as something imaginary or made up, 'fictitious' is the accurate term, despite its slightly pejorative connotation, as the term 'fictive' has a broader usage and can be used for both works of fiction and the events within them.

We have seen that fiction may come in different guises, and that, in the debate around the paradox, it is mostly the nonexistence of the intentional object that is highlighted. However, I will focus on the presentational aspect of fiction – that is, the circumstance that fictions are mediated through a form of presentation. Chapter 3 begins with the insight that we always encounter fiction through representations; consequently, my attention will shift from works of fiction to representations in general, and more specifically, to the intentionality of representations. However, in order to avoid confusing 'representations', in the sense of 'works with representational content', with the more abstract notion of representation, as used, for instance, by Searle – where it denotes a property of mental states shared with speech acts[31] – I will refer to 'physical items or performances with representational content' as 'artefacts' instead. I will not use the term 'artefacts' as synonymous with works of art, as the latter are not always representational – and only if representation is involved can we speak of it in terms of fiction. The term will also apply to representations that are about factual states of affairs, such as newspaper content. As opposed to 'representations', the term 'artefacts' emphasises physical items or events and the fact that they are made or realised by someone. This distinction becomes important when I point out the presentational aspect that all artefacts share, and that we have access to represented content only through its form of presentation.

Finally, it seems necessary to qualify some of the uses of the term 'intentionality'. Although I will address different understandings of intentionality in more depth in chapter 4, the landscape of the term's varying uses, types, and the adjectives applied to it, can be rather confusing. It therefore seems appropriate already to introduce at this stage some of the variations in which the term is used, and which will be mentioned throughout this study. Generally, intentionality is understood as the capacity of the mind, or the property of mental states, to be directed towards or to be 'about' something. Different types of intentionality have emerged, such as *derived intentionality*, *affective intentionality*, *phenomenal intentionality*, and others. These notions typically either apply intentionality to a specific domain or specify how it is to be understood with

31 'The main function which language derives from intentionality is, obviously, its capacity to represent' (Searle 1983, 175).

respect to a particular subject. In addition to these various types of intentionality, the term is also further specified through adjectives or paraphrases, to emphasise a certain interpretation or use of the concept. *Propositional intentionality*, for instance, conceives of intentionality as the 'having of a propositional attitude': as the property of certain mental states to be directed towards propositional content under a specific mode of presentation. Beliefs, for instance, are characterised by a commitment to the truth of the proposition (mind-to-world direction of fit), whereas desires are oriented towards the realisation or fulfilment of the state of affairs expressed by the proposition (world-to-mind direction of fit). Propositional content is typically articulated through *that*-clauses, which designate the object or state of affairs intended by the mental attitude in question (e.g., 'I believe *that* it is raining', 'I hope *that* it will rain tomorrow'). Instances of *non-propositional intentionality* include perception – such as seeing a red apple – which is directed not necessarily at a proposition but at a perceptual object, as well as affective states such as anxiety or joy, which may lack clear propositional content. Furthermore, some mental states are difficult to grasp in terms of propositional intentionality, either because, although they have an object, they are not formulated using a *that*-clause (e.g., 'I pity Anna'), or because they involve a complex, affective, and embodied relation to the world that is not easily reducible to discrete propositional content. Moreover, in propositional attitudes, content is typically subject to truth or falsity. However, not all mental states aim at truth on these terms. For instance, 'I regret that I said that' is not falsifiable in the same sense as a belief; rather, it expresses an emotional orientation towards a past action. To avoid a strictly propositional understanding of intentionality, it is sometimes described as the 'having of an intentional object' or referred to as *directedness*. The former does not require propositional content, that is, to be figured in a sentence that can be true or false, but only the presentation of an object the mental state is 'about'. I will refer to this as *objectual intentionality*. The latter term, *directedness*, further allows for instances of intentionality that do not involve a determinate intentional object, such as certain moods. Some authors also use the terms *aboutness*, or *of-ness*, whereby a propositional understanding is often assumed, but more generally the terms are understood as referring to a mental state 'having content'. Furthermore, the term *mindedness* has emerged as a way of referring to what it means for a creature or system to be capable of intentional states and is often used in contrast with 'mindless' or purely mechanical systems. The term is typically used to emphasise the *capacity* to have intentional states. Mindedness is not synonymous with intentionality; rather, it concerns the broader capacity of a system (human or artificial) to possess mental properties, including intentionality, consciousness, normativity, and so forth. There are many further uses, nuances, and tendencies in the understanding of the term, but the ones outlined here will suffice to distinguish the different notions of intentionality employed by the authors discussed in this study.

This paragraph is intended to provide only an initial orientation. Clarifying the use of the term 'intentionality' is the central objective of this study. Accordingly, I will pay close attention to and bring to light the implicit understandings present in the different authors discussed throughout. Chapter 4 is dedicated to examining the discrepancies found in the explicit and implicit understandings and uses of the term. There, I analyse in depth the paraphrases of intentionality as *content* and *directedness;* and distinguish between intentionality as a *property* of mental states and as an *activity* of the mind.

2. Dealing with emotions

> By degree I made a discovery of still greater moment. I found that these people possessed a method of communicating their experience and feelings to one another by articulate sounds. I perceived that the words they spoke sometimes produced pleasure or pain, smiles or sadness, in the minds and countenances of the hearers. This was indeed a godlike science, and I ardently desired to become acquainted with it.
> *Mary Shelley, Frankenstein; or The Modern Prometheus*

Our emotions, like our thoughts, are mental states which are often formulated as directed to an object or state of affairs – our pity is *for*, our fear is *of* something – and they are also experienced as being directed towards the respective object or state of affairs. However, there is an intuition that it matters differently to us if the objects of our emotions exist, or if the objects of our thoughts do. Emotions seem to be situated closer to action: we want to do something about them. There is a motivational force inherent in emotions, which is not necessarily present in thoughts. This becomes apparent in our emotional responses to fiction: having beliefs aboutFrankenstein's monster – how, for instance, it was created out of inanimate matter – is one thing; being afraid of Frankenstein's monster is another. Both entertain the possibility that Frankenstein's monster exists, but the latter seems to imply that we can, or should, do something about it.

The paradox of fiction asks us to account for the intentionality of emotions. Emotions, like thoughts, are directed to, are 'about' things. Is this the same kind of directedness? We can distinguish two points here: on the one hand, intentionality is not necessarily attributed to all emotions. Some say that not all emotions are 'about' something – startles or depression (in different ways) do not necessarily have anything as their object. On the other hand, having an emotion seems to have different implications than having a thought, namely, that emotions typically come with tendencies or dispositions to act. Both points are related to proposition (2) of the paradox: that to experience emotional responses towards an object we must believe in the existence of that object. Only few theorists would still agree with this claim, namely, proponents of a narrow cognitive theory of emotions. Most believe not that existential belief is a necessary component for emotion, but that something less than a belief (like a thought or imagination) suffices. This shift in the concept of emotion solved the doxastic contradiction of the paradox – that we cannot believe and not believe p at the same time. However, there are still practical concerns to consider. Something about an emotional reaction towards a fictional object seems to be different from a thought about it. One of the main arguments is that Ref have a reduced behavioural output: we might be afraid at the sudden appearance of danger in the film, but we typically do not flee the cinema in response to it. The shift in emotion theory helped to clarify the doxastic contradiction of the paradox. Questioning the concept of intentionality, I believe, can help us solve issues of what Vendrell Ferran (2014) refers to as practical rationality.

In this chapter, I want to address the first of the two perspectives I proposed for examining intentionality through emotional responses to fiction: (A)' whether there is a difference between the intentionality of thought and the intentionality of emotion. I begin by presenting the argument of reduced behavioural output of Ref, discussing a frequently cited set of examples by Kendall Walton, with particular attention to his understanding of intentionality. I then outline the role intentionality has played, and continues to play, in our understanding of emotions and present the concept of affective intentionality. Proponents of affective intentionality, such as Bennett Helm, Sabine Döring, and Jan Slaby, argue that emotions are intentional states sui generis, and that their intentionality must be treated differently from that of beliefs and desires. I then apply the concept of affective intentionality to emotional responses to fiction, and examine whether it can help us account for the reduced behavioural output of Ref. Finally, I point out an implicit, but significant, difference in the general understanding of intentionality held by proponents of affective intentionality.

2.1 Kendall Walton's three-men example

The paradox of fiction arose from the intuition that our emotional responses to fiction differ in significant ways from our emotional responses to real situations, and that there seems to be something incoherent about having a 'fully-fledged' emotion towards a fictional character or state of affairs. In *Mimesis as Make-Believe* (1990), Kendall Walton describes our connection to fictional worlds as 'strangely schizophrenic': physical interaction is blocked, whereas psychological interaction is not. We can pity Anna Karenina, but we cannot do anything for her; we can be afraid of Frankenstein's monster, but we do not flee the theatre. Walton develops this idea through a sequence of examples featuring three men – Henry, Charles, and Aaron – which I will briefly present, along with the argument in which they are embedded:

> Henry, a backwoods villager watching a theatrical performance, leaps to the stage to save the heroine from the clutches of the villain and a horrible death. (1990, 192)

The actress is not in danger, only the character is, and Henry cannot save her, even if he would prevent the play from continuing. It seems not only inappropriate to interfere in a theatrical performance but also simply ineffective, at least if the goal is to save the fictional character. Although we might feel very strongly for the heroine, intervening in the performance is not a viable course of action, because fictional worlds are 'insulated' from the real world, and, as Walton puts it, 'one person can save another only if they live in the same world. *Cross-world saving* is ruled out' (1990, 195). Since the gulf between our world and the fictional world appears unbridgeable, Walton questions whether we truly have psychological attitudes towards fictions, or whether they are merely construed as such:

> Charles is watching a horror movie about a terrible green slime. He cringes in his seat as the slime oozes slowly but relentlessly over the earth, destroying everything in its path. Soon a greasy head emerges from the undulating mass, and two beady eyes fix on the camera. The slime, picking up speed, oozes on a new course straight toward the viewers. Charles emits a shriek and clutches desperately at his chair. Afterwards, still shaken, he confesses that he was 'terrified' of the slime. (196)

Although Walton grants that Charles is having a genuine emotional experience, he argues that Charles cannot truly fear *the slime*, because he does not believe he is in danger. Based on theories of emotion that describe it in terms of beliefs and judgements, Walton maintains that Charles's description of having been 'terrified' of the slime should not be taken literally. The symptoms he exhibits, he argues, reflect a 'physiological-psychological state [of] *quasi-fear*' (ibid.). In the last example, Walton offers a comparison to a nonfictional scenario that involves irrational beliefs:

> Compare a person who hates flying. In one sense Aaron realizes that airplanes are (relatively) safe. He says, honestly, that they are and he can quote statistics to prove it. Yet he avoids traveling by air as much as possible. He is brilliant at devising excuses. If he must board a plane, he becomes nervous and upset. Perhaps Aaron believes, at a 'gut' level, that flying is dangerous, despite his 'intellectual' opinion to the contrary. And he may really be afraid of flying. (198)

Walton contrasts Aaron's case with that of Charles by focussing on the different behaviours they display. Aaron performs deliberate actions: he fears flying and devises excuses to avoid doing so. Charles, by contrast, does not leave the cinema or call the police. His physiological reactions are, according to Walton, merely 'automatic responses'. Because there is no resulting behaviour from which we could infer belief, Walton concludes that we cannot attribute to Charles the belief that he is in danger, whereas we can infer from Aaron's deliberate actions that he believes he is. This belief may be irrational, based on a gut feeling; nevertheless, Aaron's actions are a reasonable consequence of his fear of flying. Charles's case, however, lacks that behavioural dimension. Therefore, Walton describes Charles's response as a 'quasi-emotion'. He does not question the genuineness of Charles's emotional experience but rather its 'toward': that is, whether the emotion is truly *about the slime*. 'To take him literally and straightforwardly when he testifies to being afraid of the slime is dangerously presumptive' (1990, 200), Walton claims, comparing the case to children shouting, 'There is a monster in the basement!' during a game of make-believe.

Walton construes and evaluates these examples by their outcomes – that is, by the actions that follow from, or are motivated by, the fictional or factual events. Both Henry's and Charles's emotional reactions are, according to Walton, 'genuine experiences'. However, they are not acted upon. It is not a real option to run onto the stage in order to save the heroine, because we cannot *actually* save her.Even if we were to change the script, this would result in a new work rather than alter her destiny.[1] And although Charles displays strong physiological symptoms, Walton considers these to be automatic responses rather than deliberate behaviour. Only Aaron truly acts upon his emotions – albeit against his better judgement. Since Walton emphasizes the behavioural outcome of the three scenarios, and categorises the different forms of emotional engagement accordingly, the main claim he is trying to establish through these three examples is that reduced behavioural output is the distinguishing criterion for emotional responses to fiction.

Distinguishing Ref from other emotions on the basis that they do not motivate action has been proposed by several authors (cf. Lamarque, 1981; Carroll, 1990; Currie, 1991; Yanal, 1999), thoughit has also been challenged in various ways. Vendrell Ferran summarises the debate as follows: first, not all emotions motivate action; second, it cannot be said that Ref never do (2014, 332–333). On the one hand, we can call into question whether reduced behavioural output is a distinctive criterion for Ref, since many other emotions also fail to imply direct action or observable behaviour. Moran (1994) has pointed out that Walton places too much emphasis on the aspect of fictionality in his theory. He suggests we should instead compare Ref to other cases involving nonactual scenarios: for instance, 'things that might have happened to us but didn't, things we might have done, how things could have turned out differently' (77–78). Moran claims that Ref are not the only emotions that are characterized by reduced behavioural output, and indeed we can easily think of several other examples of this sort.

On the other hand, we can also think of cases in which Ref do lead to action, while Walton's examples are limited to cases in which no *direct* (re)action is involved. Along these lines, Christiana Werner (2020) argues that emotional responses to fiction can lead to various forms of action. She specifies that only 'immediate object(f)-related actions' – that is, actions involving physical interactions with the fictional object – are typically absent (9–10). We can, for instance, genuinely admire Superman, and our behaviour may be motivated by a desire to be like him; we simply cannot shake his hand.Since reduced be-

1 'It is arguable, however, that altering a painting or novel does not affect what happens in the fictional world but rather creates a new work and a new fictional world to go with it' (Walton 1990, 193n2).

havioural output is neither exclusive for Ref (per Moran), nor the characteristic for all Ref (per Werner), it cannot serve as a distinction criterion.[2]

Still, reduced behavioural output does appear to be a feature of some Ref. So how can we account for it? To explain this lack of action, Werner turns to the doxastic conditions of Ref. She argues that certain actions are absent in Ref due to the absence of object(f)-directed intentions. This, in turn, results from the specific doxastic context in which we engage with fiction: 'typical readers and viewers of fiction lack the belief that they can perform object(f)-directed actions' (2020, 13). This explains the reduced behavioural output of Ref without requiring that they 'differ in any significant way from non-fictional emotions' (ibid.). The difference lies not in the emotion itself but in the unmet doxastic conditions for action-oriented intentions.

I want to approach the issue from the point of view of intentionality. Other than explaining the reduced behavioural output of Ref by questioning the status of the emotion (as Walton does), or by identifying different doxastic conditions for action (as Werner suggests), I want to focus on the 'towards' – that is, the directedness or aboutness of the mental states involved. Is the 'towards' of our admiration for Superman essentially different from the 'towards' of our thought that he can fly? Put differently: is the intentionality of an emotion somehow different from the intentionality of a thought? I will begin by revisiting Walton's examples from the perspective of intentionality and analyse how intentionality is understood in his account. I will then consider the role of intentionality in emotion theory in general and introduce the concept of affective intentionality in particular.

2.1.1 The role of intentionality in Walton's argument

In his examples of Henry, Charles, and Aaron, Walton highlights the reduced behavioural output of Ref. He uses behavioural output as an indicator for the belief implied by, or accompanying, the emotional state. This allows him to infer that Charles does not really believe he is in danger, based on his automatic responses and reduced behaviour. In a broader sense, the reduced behavioural output of Ref serves Walton in preparing his participatory account of fictionality, in which 'representations are props for appreciators to engage in games of

2 Goldie (2009, 232), for instance, also argues that motivational force is not a constitutive criterion for emotion, maintaining that while emotions often motivate action, this is not a necessary feature.

make-believe'.[3] The context in which he presents these examples is intimately tied to intentionality, and how he discusses them also reveals his understanding of that concept. I will take a closer look at the argumentation surrounding the examples of Henry, Charles, and Aaron in order (1) to show that intentionality is one of Walton's main concerns in this context and (2) to analyse his use and understanding of intentionality.

The central claim Walton upholds in presenting these examples is that, although fictions can induce genuine emotional responses, fictional characters and situations are not the objects of these emotions. He believes that people do not literally pity or grieve fictional characters.[4] This does not question the genuineness of the emotional experience of Ref; his notion of quasi-emotion should not be misunderstood as implying that Ref are less intense or otherwise phenomenologically different.[5] Rather, as I will show, he is concerned with the intentionality of our emotional responses to fiction. He refers to aspects of intentionality immediately before and after introducing Charles's case:

> We do indeed get 'caught up' in stories; we frequently become 'emotionally involved' when we read novels or watch plays or films. But to construe this involvement as consisting of our having psychological attitudes towards fictional entities is to tolerate mystery and court confusion. (1990, 96, my emphasis)

> Was he terrified of it? I think not. Granted, Charles's condition is similar in certain obvious respects to that of a person frightened of a pending real-world disaster. His muscles are tensed, he clutches his chair, his pulse quickens, his adrenaline flows. Let us call this physiological-psychological state 'quasi-fear.' But it alone does not constitute genuine fear. (ibid., my emphasis)

---

3 Cf. Walton (1990, 158), chapter 5, section 3, and chapter 6..

4 Cf. Walton (1990, 202).

5 Against the argument of *suspension of disbelief* or 'half-belief', Walton contends that 'the symptoms that Charles does exhibit are not symptoms of mere suspicion that the slime is real and a queasy feeling of half fear. They are symptoms of the certainty of grave and immediate danger and sheer terror [...] that Charles half believes he is in danger and is half afraid, less than justice to the intensity of his [Charles's] reaction' (1990, 198). He emphasizes the phenomenal genuineness of these quasi-emotions throughout the book.

In the sentence preceding the example of Charles, Walton expresses his scepticism towards the idea that our emotional involvement with fiction consists in psychological attitudes directed towards fictional objects. In the sentence following the example of Charles, he then questions the intentional object of Charles' fear. Walton does not deny that Charles was frightened – the phenomenology of Ref is 'granted' (ibid.) – but rather that the emotion was fear *of the slime*. These framing statements clearly show that Walton's considerations revolve around issues of the intentionality of Ref.

So how does he qualify this simple 'I think not': his rejection of the idea that Charles fears the slime? Walton takes three things as given concerning Ref: that they exhibit reduced behavioural output, that they are genuine emotional experiences, and that they are typically described as being 'of' or 'about' fictional characters and situations. Among these, he identifies the third as the weak point.

He first considers a claim from what he calls 'standard theories of emotion', which we would now call narrow cognitivist theories of emotion: that 'fear necessarily involves a belief or judgment that the feared object poses a threat' (1990, 197). On this basis, Walton argues that Charles does not genuinely fear *the slime*, because he does not believe that *it* poses a threat. And Charles does not believe that the slime is a threat, because it does not really exist. Walton then considers objections to this cognitivist claim by suggesting cases where a cognitive component, less than a belief or judgment, may suffice for emotion. However, he maintains that such cases differ from Ref due to the presence of deliberate action: in Ref, automatic responses, such as a racing heart or tensed muscles, do not prompt actions, and thus do not support the claim that the person truly believes the object is a threat.[6] Furthermore, he suggests that behavioural

6 Walton compares the examples of Aaron and Frances in this context: Aaron is afraid of flying, and although he realizes that planes are relatively safe, he avoids traveling by air. Frances (an example by Patricia Greenspan, 1981) was attacked by a rabid dog, and now she is afraid even of old, lovable Fido, although she knows that he had his rabies shot and is practically toothless. She avoids Fido while being happy to let others play with him (1990, 201f). The two cases differ in the cognitive component involved: Aaron may believe at a 'gut-level' that flying is dangerous, while it would be awkward to attribute the belief that Fido is dangerous to Frances. However, both cases involve deliberate actions. These are contrasted by Walton with the absence of action in Charles's case. In opposing Charles to Aaron, Walton emphasizes that automatic responses such as sweating or high pulse do not require reasons, only causes (cf. 1990, 198f). In Frances's case, there are no beliefs that explain her actions, yet she still ex-

output is more relevant to the assessment of emotion than the cognitive component involved in emotion. Reduced behavioural output becomes, for him, a key reason to reject the belief component in Ref. However, as we saw in the last chapter, behavioural output cannot be taken as a distinction criterion for Ref.

Walton then considers the possibility that our Ref might be directed at something other than the fictional characters and situations themselves (the so-called *surrogate-object-solution*): perhaps our emotion is directed towards the performance, the description of the fictional character or state of affairs, or towards real people who are 'like' the fictional character. Lamarque (1981), for example, argues that our fear is directed, not at the fictional character, but at the thought or description of that character. Walton rejects this for two main reasons. First, even from a cognitivist perspective, this move offers no advantage, since Charles neither believes that the slime exists nor thinks that the description of the slime is dangerous. Second, Walton invokes the phenomenology of the emotional experience:

> Moreover, his experience simply does not feel like fear of a thought; characterizing it as such flies in the face of its phenomenology. And it is the slime, not the thought, that Charles so inevitably and unabashedly describes himself as afraid of. (1990, 203)

A similar argument is given against the idea that our emotion is directed at a general type: for instance, that our pity for Anna is really pity for people like her, or for people in like situations. Again, Walton objects on similar grounds: 'it is no accident that we speak of sympathizing with or grieving for *Anna*' (1990, 204). We do not only formulate Ref as directed at particular fictional characters; we also experience them that way. He allows only one exception: that we might be afraid of the performance itself. He alters Charles's case to that of an elderly man with a heart condition: this 'Old Charles' might fear the movie because it could cause him a heart attack. In this case, the emotion is directed not towards the slime, but towards the depiction of the slime 'as being especially aggressive or threatening' (1990, 202) and involves the possibility of real physical harm caused by such a depiction.

---

hibits characteristic fear behaviour. Charles, by contrast, shows no tendencies to act (running away, calling the police, etc.). Walton concludes that 'fear is motivating in distinctive ways, whether or not its motivational force is attributed to cognitive elements in it' (1990, 201).

What, then, distinguishes Walton's rejection of Charles's fear as being directed *at* the slime? The argument can be summarized as follows: Charles does not display action tendencies *directly related to* the slime – for example, he does not run out of the cinema – because he does not believe that the slime poses a real threat to him. He does not believe it to be a real threat because he knows the slime is fictional. And he assumes that fearing something necessarily involves believing oneself to be endangered *by* that thing. Thus, Walton concludes that we cannot have genuine emotional experiences *towards* fictional objects. Charles's quasi-fear is a genuine emotional experience, but it is not fear *of* the slime. And yet, we do describe and experience Ref as directed towards fictional characters and situations. Walton insists that neither a half-belief nor a momentary confusion, nor a surrogate object can account for this. Furthermore, he dismisses different challenges to emotions necessarily involving a judgement or belief. His solution is to propose a new account of fictionality – one that accommodates how we describe Ref as being directed towards fiction, and that respects the phenomenology of experiencing emotions as directed towards fictional characters and situations. He presents a 'positive account': that of wilful participation in games of make-believe. According to this model, we have quasi-emotions operating with quasi-beliefs about fictional truths, which enables fictional characters to serve as the intentional objects of Ref.

Charles is quasi-afraid as a result of realizing that fictionally the slime threatens him is what generates the truth that fictionally he is afraid of the slime. (1990, 245)

Through his wilful participation in games of make-believe, Charles becomes quasi-afraid of the slime because the representation of the slime generates quasi-beliefs about it.[7] Walton understands propositions about

7 He does not use the term 'quasi-beliefs' but talks about 'fictional truths'. I would argue, however, that it is illuminating to use the term in analogy to 'quasi-emotions'. Doing so does not diminish the concept of 'belief', just as 'quasi-emotion' does not call into question the intensity or genuineness of the phenomenal experience of the emotion. As Walton uses it, the prefix *quasi-* merely indicates fictionality, i.e. that something is made fictional, rather than non-existence. This is a central aspect of his theory. As he writes: 'It is clear that there is something special about fictionality, as contrasted with other intentional properties. [...] We somehow regard what only fictionally exists as being real and what only fictionally occurs as actually taking place, whereas we comfortably take what is merely believed or desired or said to exist or occur to be just that. The psychological bond we feel to fictions is a dramatic symptom of this inclination' (Walton 1990, 205). The notion of 'quasi-belief' thus indicates that fictional content

fictional objects *de dicto*, as fictional truths. In this way, Ref can be directed towards fictional objects because the intentional relation is not 'between worlds'. However, Charles's fear can only be directed towards a fictional object because he is 'part of the fictional world of his game[, and] fictional worlds can seem to us almost as "real" as the real world' (1990, 249). This not only reinforces the separation between the 'real-world' and 'worlds-of-make-believe' – a distinction Walton maintains throughout his work – but also takes it a step further:it introduces a 'separation between our actual mental lives and the mental lives we lead in the worlds of our games of make-believe'.[8]

Examining the context of the examples of Henry, Charles, and Aaron confirms that concerns about intentionality are central to Walton's project. I would go as far as to argue that issues of intentionality figured as one of the main reasons for Walton to develop his theory of make-believe. However, the concept of intentionality itself was not problematised by him, as it was generally neglected in the debates around the paradox. Although he does not use the term 'intentionality', but various other descriptions, it became clear that issues of intentionality are at the centre of his reasoning.

2.1.2 Walton's understanding of intentionality

While Walton's argumentation revolves around intentionality – more precisely, around the intentionality of emotions – his understanding of intentionality remains implicit. I will attempt to trace this implicit understanding here. At the beginning of his second chapter, which, after outlining the theoretical questions around representations, addresses their appreciation, Walton asks: 'What relations hold between the real world and the fictional world?' (1990, 191). He continues to describe these relations as 'strangely schizophrenic': on the one hand, we cannot physically interact with fictional worlds and their contents. 'On the other hand, we seem to be in psychological contact with characters, sometimes even intimate with them' (ibid.).

He continues to describe this 'psychological contact' with fictional characters as a 'psychological link' (ibid.), 'psychological interaction' (1990, 192), 'psy-

---

functions as a kind of truth within the fictional framework: Anna is a fictional character rather than non-existent.

8 Walton (1990, 252). Although Walton admits that these different mental lives overlap in the case of fictional characters based on real people, such as Ivan the Terrible or Caesar, he insists that they are strictly separate for purely fictional objects.

chological bond' (1990, 196), or even as seeming 'intimacy'[9] with them. Furthermore, he consistently uses the term 'psychological attitudes' to describe the intentionality of our emotional responses in general, and to distinguish the intentionality of fiction-directed emotional responses in particular:

> To allow that mere fictions are objects of our psychological attitudes while disallowing the possibility of physical interaction severs the normal links between the physical and the psychological. (1990, 196)

> But to construe this involvement as consisting of our having psychological attitudes toward fictional entities is to tolerate mystery and court confusion. (ibid.)

He only uses the term 'intentional' when he distinguishes fictionality from other 'intentional properties of propositions' (1990, 205), claiming that the comparison of fictionality with

> being believed or desired or claimed fails to bring out the difference that seems to obtain between physical relations across worlds and psychological ones, the 'impression we have [...] that things known to be merely fictional are objects of our psychological attitudes' (ibid.)

Both the analogy between psychological attitudes and propositional attitudes, and the comparison of fictionality to other 'intentional properties' of propositions, suggest that he understands intentionality according to a linguistic model. Although he does not state this explicitly, I believe that his understanding of intentionality is basically that of linguistic reference. This is, in addition to the points I have already mentioned, further confirmed by his treatment of the objects of representation.

In an earlier chapter, in which he delineates his notion of representation, he criticizes the comparison often made between representations and their objects and that of linguistic reference. Arguing against Goodman, Walton

9 Walton (1990, 196, 205) uses the term 'intimacy' to emphasise the phenomenal dimension of our relation to fictional characters. However, he later claims that his theory of make-believe 'accounts for the apparent asymmetry in our relations with fictional characters, the fact that we seem psychologically intimate with them but physically cut off from them. There is not really such intimacy. It is only fictional, not true, that we feel for Willy Loman or detest Iago'. (1990, 241)

opposes this understanding of representation as denoting, because, in his view, there is a fundamental difference between representing and what he calls 'matching' (1990, 122). His main claim is that objects – that is, actual existing objects – are inessential for representations, while they are essential for denoting in linguistic referring expressions. Referring expressions are, according to Walton, to be understood in terms of their denoting roles, while representations are not. He contrasts the case of non-denoting referring expressions, such as 'the only person to climb Mount Everest', or 'Grindel', to non-denoting representations: for example, a drawing of dinosaur footprints discovered in Arizona, which actually are not the footprints of a dinosaur but marks made by a frustrated palaeontologist. In this case, the representation would not have an actual object. However, according to Walton, representations do not have to be denotative in this sense: 'Drawing a bison is thought of always as *creating, producing* an (imaginary) bison' (1990, 125). He concludes that 'the notion of the representation is thus independent of that of denoting', and that 'the notion of objects of representation is inessential to that of the representational' (ibid.).[10] He sees an analogy between predicates and representations: both pick out, or specify, properties. However, according to Walton, while the dinosaur-picture specifies the proposition that there is a dinosaur, it does not attribute the property it picks out to a particular thing. There are cases in which we don't know whether a represented object exists or existed, that is, whether the work has an actual object or not:

> We may have no idea whether or not an aboriginal race painting or a contemporary sketch portrays a real person. But this uncertainty may scarcely matter; our experience need not depend in any fundamental way on which we suppose to be the case. (1990, 132)

His argument that the actual existence of objects is inessential to representation contrasts with his belief in their essential role in the denoting function oflinguisticreference.

Of course, he is talking about representations here in the sense of works of fiction – like novels, paintings, or plays – not about mental states. How-

10 Walton argues against understanding the relation between representations and both fictional and actual objects on the model of linguistic reference, emphasizing instead the 'creative' function of representations: 'representations "create" by making existential propositions not true but fictional' (1990, 127).

ever, these considerations in his first chapter inform later passages in which he compares fictionality to other intentional properties of sentences (in his fifth chapter), and where he rejects the understanding of emotions as psychological attitudes. What seems so problematic to him, in understanding Ref as psychological attitudes, is that it would require the intentional object to exist:

> But grief, as well as pity and admiration, would seem to require at the very least awareness of the existence of their objects. It is arguable that for this reason alone appreciators cannot be said actually to pity Willy or grieve for Anna or admire Superman. (1990, 204)

Instead, he chooses to construct them as quasi-emotions. We can see how this is partly determined by his adherence to narrow cognitivist theories of emotion, which understand emotions according to a linguistic model: as psychological attitudes directed towards an intentional object. Here the intentional relation is understood as analogous to linguistic reference, which requires the existence of its object. This explains why Walton is concerned about whether we should take Charles 'literally and straightforwardly when he testifies to being afraid of the slime' (1990, 200),[11] and why, according to him, we must explain Ref differently:

> We need a positive account of appreciators' experiences, a well-articulated alternative to literal-minded acceptance of ordinary claims that they "pity Willy", "grieve for Anna", "fear the slime", and so on. [...] it is more than evident already that the legitimacy of such literal mindedness is not to be taken for granted. (1990, 204)

On this and another occasion, Walton uses the word 'mindedness'. The term addresses the *capacity* to have intentional states, and it typically concerns the broader capacity of a system to possess mental properties, including intentionality, consciousness, and normativity.[12] He does not appear to use the term in

11 To emphasize that such statements should not be taken literally, Walton sometimes places these expressions in quotation marks: 'to say that someone "pities" Willy Loman or "admires" Superman'; and 'the recent literature narrowly focused on Charles's "fear" and on the experience of being "moved" by Anna Karenina' (1990, 200n8).

12 John McDowell (1994), for instance, uses the term to discuss our conceptual capacities and their role in perception, judgement, and meaning, contrasting these with mere causal responsiveness.

this broader sense but, more specifically, to indicate that a mental state or artefact has a (propositional or representational) content. More precisely, I suggest that Walton's understanding of intentionality is best described in terms of propositional intentionality: as the view that intentional mental states involve the 'having of a propositional attitude', which entails existential commitments towards the object of that proposition. This is, for instance, why he describes thoughts about fictional characters in terms of fictional truths: they make 'existential propositions not true but fictional' (1990, 127). It is also why he refuses to understand Ref as 'psychological attitudes', instead describing them as 'quasi-emotions'. Propositional intentionality is a common understanding of intentionality and is also prevalent in emotion theory. In this view, there is no difference between the intentionality of beliefs and that of emotions; rather, the former must be accounted for in emotions. According to this understanding, not all emotions exhibit intentionality, because not all emotions have a determinable object they are 'about' or 'of' – as in the case of moods. It was the standard understanding of intentionality in early cognitive emotion theory, which held that an emotion is always accompanied by a belief or desire. This understanding tends to place particular emphasis on the intentional object, to the extent that the intentional object figures as an indicator of intentionality.

Arguably, it is this understanding of propositional intentionality that leads some theorists concerned with the paradox to try to 'get around' intentionality. I have already mentioned Lamarque's (1981) proposal to substitute the object of our emotional response with something nonfictional, like a description of the fictional entity, or real people/events similar to the ones represented. Another way to avoid intentionality is to focus on reflexive affective responses and moods, which do not have a determinable object. JinheeChoi(2003), for instance, claims that fits and startles do not entail an intentional object, and are rather induced by formal properties of the film itself. He distinguishes two cases: the contributory case, in which the effect of the representational content (the narrated story) is reinforced by formal properties, and the sufficient case, where the perceptual stimulus is a sufficient cause for emotional arousal. He focusses on the latter, which he calls reflexive affective responses, to show that some emotional responses do not exhibit intentionality in the sense of 'having an intentional object'. He claims that formal properties can be causes of our emotional responses; however, they do not figure as the objects of emotion. Since there is no intentional object,there is no inconsistency in cases of Ref,

Choi concludes.[13] However, not all Ref are fits and startles: some imply complex cognitive involvement, such as, for example, admiring a fictional character. The range of our Ref cannot be reduced to emotions which do not have an intentional object. Thus, we must account for the intentionality of Ref.

Walton emphasizes that we have to make sense of why we seem to have a 'psychological bond' to fictional characters, even though we cannot physically interact with them: 'cross-world saving is ruled out, and for similar reasons so is cross-world killing, congratulating, handshaking, and so forth' (1990, 195). His strategy is to take our emotion over to the other realm, such that our quasi-emotions and quasi-beliefs participate in games of make-believe. This is due to his understanding of propositional intentionality and his reliance on the 'standard' theories of emotion at the time – namely, narrow cognitivist theories. In my analysis, it emerged that intentionality is one of the main concerns and constraints of Walton's theory.

Nowadays, following the shift from narrow to broad cognitivism, few theories would still uphold that a belief is necessary for an emotion.[14] This shift has had a major influence on the paradox. However, the inconsistencies of the paradox do not hinge only on the concept of emotion but also on that of intentionality. Even if we no longer insist that a genuine emotion requires the existence of its object, there are good reasons to attribute intentionality to emotional responses – for example, in order to rationally make sense of actions that follow from, or are part of, the emotional experience. To arrive at an adequate theory, not only must the concept of emotion be refined, but also that of intentionality. We cannot get around intentionality, and we must unearth the often implicit understandings of intentionality at play. How intentionality is used and understood must be questioned.

2.2 The intentionality of emotions

The firstof the two perspectives from which I proposed to examine the intentionality of Ref is (A)': whether the intentionality of our emotions is different from the intentionality of our thoughts. This follows the intuition that having an emotion seems to have different implications than having a thought. The

13 I will return to Choi's and Lamarque's argumentation in chapter 3, when discussing the advantages and disadvantages of applying the concept of derived intentionality to Ref.

14 Cf. Vendrell Ferran (2014).

most cited reason for this, in this context, is that emotions are somehow closer to actions: that they come with tendencies or dispositions to act. I first discussed Walton's example of reduced behavioural output, with special attention to his understanding of intentionality. Now, I want to turn to the debate about intentionality in emotion theory. Is the intentionality of emotions the same kind that is attributed to thoughts, and is it an aspect that only some emotions have; or do we have to take into consideration that emotions and thoughts involve different kinds of intentionality? In the following sections, I will first examine the role intentionality plays in emotion theory and then present an account that claims emotions have a kind of intentionality sui generis – namely, *affective intentionality*.

2.2.1 The role of intentionality in emotion theory

Intentionality, along with phenomenality, is one of the central features commonly attributed to emotional experience. Phenomenality refers to the feeling aspect of an emotion: the 'what-it-is-like' to undergo a particular emotional experience. Intentionality, by contrast, is generally understood as what provides the emotion with its 'about'. Often, it refers to the object of emotion – what or whom the emotion is directed at. We experience our emotions as being directed beyond ourselves:[15] our fear is *of* thesnake in front of us, and ourempathy is *for* a particular person. Furthermore, we also formulate emotions in this way.

Several theories about emotion emerged in the last century, with their origins largely attributed to the theories of William James (1884) and Wilhelm Wundt (1893). However, emotions only became more widely discussed in the philosophical debate with the so-called *cognitivist turn*, which focussed on the intentionality of emotions – that is, the attribution of judgement and cognition to them. Sabine Döring (2009b) points out that whereas the phenomenal, or feeling, aspect of emotions appeared somewhat isolated within the theory of mind, the evaluative or representational aspect, attributed to emotions by *cognitive theories*, opened the possibility of integrating emotions into broader discussions – such as those about rationality, values, and agency:

15 There are also self-referential emotions, usually referred to as 'reflexive emotions', such as shame or pride. However, this does not contradict the claim that emotions are directed beyond ourselves; rather, it challenges the idea that emotions would be directed at changes in our body (see below).

> Because through this [the attribution of evaluative/representational content] the possibility opens that emotions are not mere "blind motivators". Possibly, emotions can not only cause other states and actions, but also make them rational. To have a representational content means to be subject to conditions of correctness. As, for instance, fear represents a snake as being dangerous, this representation can be correct or incorrect, and, thus, the emotion appropriate or inappropriate. (Döring 2009b, 17, my translation)

Today, both phenomenality and intentionality are considered essential features of a theory of emotion, although they were initially seen as difficult to reconcile. These two aspects gave rise to two different and partly conflicting traditions in emotion theory: the *feeling theory* and the *cognitive theory*. The key difference lies in what is considered the core of emotion: whether it is a feeling/perception or a thought/judgement. In other words, the question is whether phenomenality or intentionality is seen as essential. A paradigmatic formulation of the first position is William James's well-known claim that

> the bodily changes follow directly the perception of the exciting fact, and [...] our feeling of the same changes as they occur is the emotion. (1884, 189–90)

Proponents of early feeling theories consider a feeling, or set of sensations, to constitute an emotion. For them, the perception of bodily changes *is* the emotion (thus, neo-Jamesian theories are also called *perception theories*). Feeling theories struggle to integrate intentionality into their concept of emotion. They have difficulty accounting for the fact that we experience and formulate emotions as being 'about' things in the world. We do not experience fear of a snake as directed towards bodily changes (e.g. the tensioning of our muscles or the rise of our pulse), but rather towards the snake itself. The feeling approach also has difficulties distinguishing emotions from mere feelings, because the cognitive aspect (i.e., knowing or believing something has occurred) and the evaluative aspect (i.e., judging that what occurred is dangerous or offensive) are missing. As a result, it becomes difficult to say what the experience is 'about' or 'of', and theoretically then to differentiate, for instance, the feeling of a headache from the feeling of indignation.[16] To emphasise the importance of intentionality in emotion theory, comparisons are often drawn between cognitive complex emotions and physical reactions.

---

16 Cf. Lyons (2009, 144–145).

Nowadays, even theorists working within the feeling tradition acknowledge the necessity of accounting for the intentionality of emotions. Jenefer Robinson (2009), for instance, points out:

> In emotion theory the concept of intentionality has been cited in defence of the view that emotions are or entail some kind of belief, judgement, or cognitive appraisal of the world. [...] if I am in love, there must be somebody with whom I'm in love (even if it is a fictional character in a movie); if I am angry, I am angry with somebody about something. For this very reason emotions cannot merely be bits of behavior, dispositions or tendencies to act or behave in certain ways, or pattern of bodily change. Nor can they be merely bodily feelings like itches and tickles, which are not 'about' something. (2009, 220)

Another example is Jesse Prinz (2004), a proponent of perception theory, who argues that merely representing the body as being in such-and-such a state is not sufficient to constitute an emotion. For him, intentionality is necessary because it 'renders emotions amenable to rational assessment' (2004, 54): emotions can be appropriate or inappropriate. To account for the intentionality of emotions, he proposes an 'embodied appraisal theory', attempting to reconcile the James-Lange view of emotion as bodily perception with the appraisal theory's cognitive focus. Drawing on Fred Dreske's theory of mental representation (1981) and Antonio Damasio's somatic markers (1994), Prinz develops a theory in which emotions are 'embodied appraisals', which does not aim to 'supplement embodied states with meaningful thoughts [but instead to] put meaning into our bodies' (2003, 58).[17] According to the *embodied appraisal theory*, emotions are intentional in the sense that they represent the world via changes in the body.

We have seen how intentionality plays a crucial role in understanding emotions: it is through intentionality that we can justify emotions, explain their causes, and assess their appropriateness and coherence.Cognitive theories have the advantage of distinguishing emotions from physical feelings (like a toothache) and from reflexive affective responses (like itches and startles). However, they are often criticised for neglecting the phenomenal aspect of emotion, and for failing to individuate emotions sufficiently. It is possible to perform significance attributions without experiencing them emotionally: I

17 Prinz argues that emotions are perceptions of so-called 'representational files' – patterns of somatic change that represent threats, losses, and similar phenomena by virtue of being reliably caused by them.

can judge a situation to be dangerous and take actions accordingly, and I can have the same judgement while experiencing intense fear or panic.[18] Cognitive theories often fail to explain what makes emotional responses distinct from non-emotional judgements. Without integrating the qualitative or phenomenal aspect, emotions appear insubstantial, since the characteristic feeling of the experience is not accounted for. To individuate an emotion, we need not only intentionality (to distinguish a headache from indignation), but also phenomenality (to distinguish the emotionless judgment of a situation as dangerous from the experience of panic).

Although early cognitivist theories did not deny the feeling aspect, they often deemed it irrelevant to the rationality and typology of emotions. Emotions were described in propositional terms: for instance, if a feeling of fear could not be formulated as a proposition necessarily containing an object that is considered dangerous, it would not be identified as an instantiation of the emotion fear. This recalls Charles's fear of the slime and Walton's efforts to make sense of this experience. In cognitivist frameworks, the object of emotion becomes the indicator of intentionality, because emotions were understood in propositional terms, and a psychological attitude needs an object to be directed towards. The intentional object thus helps identify a feeling or sensation as an instance of emotion. For some theorists, this means that many feelings, especially moods, do not qualify as emotions, as they lack an identifiable object. Others question how to account for cases in which the intentional object is unclear.[19]

There have been efforts to integrate the aspect of phenomenality as a component of emotion in cognitivist theories, for example by so-called *composite theories*.[20] These claim that emotions are composed of several components: an intentional cognitive state (such as a judgement or appraisal), a non-intentional feeling or body-perception (through which the phenomenal experience of theemotion was explained), and a motivating desire. According to this theory, our fear of a snake, for instance, consists in our belief that the snake is

18 This insight has been advanced by several theorists. Sabine Döring, for instance, notes that an encyclopedia article on emotion and feeling from 1967 already observed that two people might judge a snake to be dangerous – one cold-bloodedly, the other gripped by fear (cf. 2009b, 32).

19 Since Kenny (1963), a distinction has been drawn between the particular and the formal object of emotion, as noted earlier. Another relevant distinction is that between the target object and propositional object.

20 Cf. for instanceLyons (1980), Robinson (1983), de Sousa (1987).

dangerous, plus a queasy feeling in our guts, plus the desire to somehow escape the danger it poses.This account tries to do justice not only to the phenomenal aspect and the intentional object of our emotional experience but also to another aspect of intentionality problematic in emotions: the so-called direction of fit. Typically a direction of fit is attributed to intentional states,according to which beliefs have a *mind-to-world direction of fit* (the mind has to fit the world, that is, our beliefs have to be appropriate to what is the case), whereas desires have a *world-to-mind direction of fit* (the world has to fit our mind, that is, our desires express what hopefully becomes the case).[21] However, emotions seem to display both directions of fit: they incorporate a representation of a dangerous snake, and at the same time the motivating force to escape that danger. In dividing an emotion into several components, typically into an informative belief, a phenomenal feeling and a motivating desire, composite theories seem to be able to meet all challenges: to combine intentionality and phenomenality of emotions, and to show how emotions can have both directions of fit.

Among others, Peter Goldie and Bennett Helm objected to such composite theories.Goldie argues that separating an emotion into an intentional cognitive aspect and a non-intentional feeling (or experiential aspect) does not do justice to the unity of emotional experience. He calls this unity of emotional experience a *substantial event*.[22] For Goldie, 'intentionality and phenomenality are inextricably linked' (2002, 242) in the emotional experience because emotions are *substantial events* that unite different mental and material features in experience – including bodily changes and actions that follow from the emotion. According to him, the theoretical distinction of mind and body should be secondary to a 'common-sense ontology' of emotional experience as a unity.[23] Bennett Helm takes a different approach. He argues that affective intentional states cannot be identified circle-free against non-affective intentional states.

---

21 The concept originates inAnscombe's well-known shopping list example (1957, §32); however, the notion of 'direction of fit' was first introduced by Searle (1975). While Anscombe attributes it to mental states, Searle also applies it to speech acts. His book *Intentionality* (1983), in particular, helped make the notion more widespread.

22 Goldie references Crane's notion of 'substance' and develops it into the notion of events: 'a substance is an object which has a certain kind of unity, partly explained by its nature. As a parallel, an event has temporal parts, not wholly present at each moment of its duration; a *substantial* event is an event which has a certain kind of unity, partly explained by its nature' (Goldie 2002, 252n26).

23 Cf. Goldie (2002, 242).

Identifying a mental state in a non-circular way means defining it without using that same mental state (or anything that presupposes it) in the explanation. That affective intentional states cannot be identified circle-free against non-affective mental states Helm calls the *problem of emotionality*: where should the specifically emotional aspect of an experience come from if its core – the intentional relation with something in the world – is not already emotional or affective? What could make an in-itself affect-free belief or an emotionless judgement into an emotional state? All candidates – such as bodily perceptions, motivations, or the intensity with which a conviction is held to be true – already make reference to emotions, and thus the explanation becomes circular. Helm claims that only if we take emotions to be intentional states sui generis, can we solve the problem of emotionality.[24] The *sui generis approach* treats the intentionality of emotions separately from the intentionality of other mental states. It claims that the way in which emotions relate to the world is so different from that of a judgement, for instance, that explanations of their intentionality have so far remained reductive. This approach proposes the concept of *affective intentionality*.

We can conclude that intentionality plays a role for emotion theory on different levels: in a direct sense, it provides the emotional experience with its 'about', and it makes emotions amenable to rational assessment. Further, on the level of definition, it is used to differentiate intentional emotions from non-intentional feelings. And historically speaking, it helped integrate emotions into wider philosophical debates.

Furthermore, the phenomenal aspect of emotion turns out to be not alone in the difficulty it presents to our attempts to integrate it into a theory of mind. Particularities of the intentionality seemingly proper to emotions are also a challenge to our understanding. Emotions seem to have two directions of fit: they are both cognitive and conative.[25]

2.2.2 Affective intentionality

Concerning our question of whether there is a difference between the intentionality of an emotion and the intentionality of a thought, at first glance it seems that proponents of affective intentionality would answer in the affirmative. Today, most emotion theories – even proponents of feeling or perceptual

24 Cf. Helm (2001, 38–41).

25 Cf. Döring (2009b, 22–23).

theories – consider it necessary to account for intentionality. Yet intentionality is often treated as something provided by a cognitive component, such as a belief, desire, or evaluation which, combined with other components like bodily changes and phenomenal aspects, composes the emotion. Proponents of affective intentionality reject this compositional view and do not treat intentionality separately from the phenomenal experience of having the emotion. Rather, in their view, our emotional experience essentially unites phenomenality and intentionality: we perceive something *as* pleasant or painful and thereby evaluate an object or situation through our perception of it.

This recent tendency reformulates emotions as intentional feelings (Slaby), felt evaluations (Helm), affective perceptions (Döring), or feelings towards (Goldie). A key claim is that the intentionality of emotions must be understood and analysed differently from the intentionality of beliefs and desires. These accounts use the term *affective intentionality* to distinguish themselves from other emotion theories, emphasizing their distinct understanding of intentionality.[26] The fundamental difference in the concept of affective intentionality and its sui generis approach lies in the *unification view* of content and quality of an emotion. The unification view holds that the intentionality and phenomenality of an emotion are inseparably united in the emotional experience:

> In an emotion, an aspect of the world is experienced as significant in a specific respect: in fear, something is experienced as a threat; in anger, something is experienced as an offence; in shame, something about oneself is experienced as a defect (according to social or moral standards); in grief, something is experienced as an irrevocable loss; and so on. (Slaby 2008a, 431)

The experience is not neutrally represented and then, as a second step, appraised or emotionally reacted upon. Instead, the phenomenality of emotions and their intentionality are inseparable. Proponents of affective intentionality naturally oppose composite theories: in such theories, the queasy feeling in our guts upon seeing a snake is treated as just another component and, according

26 Such are Stocker (1983), Goldie (2000, 2002, 2009), Helm (2000, 2001, 2002, 2009, 2011), Ratcliffe (2005), Döring (2007, 2009a, 2009b), Slaby (2008a, 2008b, 2011, 2012), Montague (2009). Furthermore, Vendrell Ferran (2016) considers also positions from early phenomenology as proponents of affective intentionality.

to its critics, as secondary. More broadly, they criticize early cognitive theories for their over-cognitivistic depiction of emotions, which, in their view, reduces emotions essentially to propositional attitudes.[27] Furthermore, affective intentional states cannot be identified circle-free against non-affective intentional states, as Helm argues.

However, positions subsumed under affective intentionality are not homogeneous: they all reject composite theories and agree on the unification view but differ in their arguments (some from a more analytic background, others grounded in the phenomenological tradition) and in what exactly they claim. There are also different conceptions of emotion between those proposed by proponents of affective intentionality. For these reasons it seems appropriate to discuss different accounts of affective intentionality. In the following, I will first briefly present the accounts of Bennett Helm, Sabine Döring, and Jan Slaby, with a focus on their respective concept of intentionality. I will then apply the concept of affective intentionality to emotional responses to fiction.

Bennett Helm

One of the main proponents of affective intentionality is Bennett Helm.[28] While the term 'affective intentionality' appears explicitly only in papers published since 2009, the concept itself was substantially developed earlier, notably in *Emotional Reason* (2001) and in the papers 'Emotional Reason' (2000) and 'Felt Evaluations' (2002). Connected with the *problem of emotionality* already mentioned, one can discern three points of critique he raises concerning the recent conception of emotions and their intentionality: the priority of intentional states to emotions, the impossibility of individuating emotions on the basis of their phenomenology, and the concept of direction of fit which, in his opinion, does not apply to emotions.

Firstly, Helm points out that according to composite theories the intentionality of emotions is understood to be dependent on the intentionality of the constitutive beliefs and desires composing the emotion. The affective nature of the emotion is explained by packaging those intentional states with a feeling aspect. This fails to capture the distinctively affective character of emotions, and he criticizes the notion that an intentional belief or evaluation must precede the emotion itself. He thinks that it is central to an understanding of

27 Consideration by Slaby, Stephan, Walter, Walter (2011) in the foreword to their jointly edited book *Affektive Intentionalität*.

28 Cf. Helm 2000, 2001, 2002, 2009, 2011, 2020.

emotions that we do not see these aspects as separately following one another, but that 'our felt awareness of emotional objects as good or bad […] just is a matter of our being emotionally pleased or pained by these objects' (2011, 72). This consideration leads him to a concept of emotions as 'felt evaluations'.

To like or dislike something, he claims, is a spontaneous evaluation, something felt rather than thought, and which can be located in the concepts of pleasure and pain. In this aspect he sees the key to the intentionality of emotions. According to Helm, emotions are species of pleasure and pain – they are conscious states with a qualitative (so-called hedonic) character.

> [E]motions do not merely involve some pleasant or painful sensation among other components, as cognitivist theories require. Rather, they are pleasures and pains and can be redescribed as such: to be afraid is to be pained by danger (and not by one's stomach); such pain is not a component of, but is rather identical with, one's fear. This means that emotional pleasures and pains, namely what one feels in having the emotion, are essentially intentional and evaluative, a sense of how things are going – whether well or poorly. (2002, 28)

Different to composite theories, in which the evaluation or belief is understood to be independent of the feeling aspect, these felt evaluations are inseparable from the feeling itself. This has been also referred to as the 'unification view' of content and quality. It is a shared point between proponents of affective intentionality; however, they provide differing explanations of it.

Helm's understanding of emotions as 'felt evaluations' is not to be confused with neo-Jamesian theories of emotion, also referred to as feeling or perceptual theories of emotion. He distances himself from these because, in his view, it is not possible to sufficiently individuate emotions based on bodily sensations. Anger and resentment, for example, are closely related emotions, but their difference is 'not intelligible merely in terms of bodily sensations' (Helm 2020, 227). Neither can weidentify affective states circle-free against non-affective intentional states, which is the *problem ofemotionality* mentioned earlier. Where should the specifically emotional nature of an experience come from if its intentional relation with something in the world is not in itself already emotional or affective? Helm claims that only if we take emotions to be intentional states sui generis, can we solve the *problem ofemotionality*.[29] In Helm's

29 Cf. Helm (2001, 38–41).

conception, emotions are feelings of pleasure and pain, and are, thus, essentially evaluations of the objects they are directed toward.The phenomenal aspect of emotions cannot be disassociated from their evaluative character, because pleasurable or painful feelings essentially bring an evaluation of their object with them: the evaluation is entailed in the feeling. To feel fear is 'to be pained by danger'; to feel satisfaction is 'to be pleased by success'. Such feelings essentially unite in their experience intentional, evaluative, and affective aspects.

However, emotions can be understood as felt evaluations only because, for Helm, emotions are commitments to the 'import' of their focusses. By 'import' Helm means 'one's caring about something' (2011, 83). And it is 'the import [that] impresses itself on us by not only grabbing and holding our attention but also priming us to act appropriately' (2011, 80). The concept of import is crucial for Helm's understanding of the intentionality of emotions. He applies it additionally to several other kinds of (in emotion theory) widely recognized intentional objects, such as the target of an emotion (that which the emotion is intuitively directed at), or the formal object of an emotion (that which determines the characteristic type of emotion). The notion of import emphasises that it matters not only if something really is dangerous but also if what is in danger matters to us. In the example of being confronted with a snake, this is not immediately obvious. The target object is the snake, and the formal object is its dangerousness, so the only question is whether it is a poisonous snake, and if we are thus justified in our fear. What Helm calls import is implicit here, because clearly our own health and well-being matter to us. However, if we consider a different example, his point might become clearer. He proposes we imagine some children playing in the street, and that we might be afraid that they damage the parked cars. In this case, the children are the target of our fear, and the formal object is dangerousness, but only if the cars matter to us does this become the focus of our attention. Otherwise, we might just perceive the children playing as part of the atmosphere of a pleasant afternoon, without giving them any particular attention, or we might even feel inspired to join their game. It is the import something has for us which makes it a potential focus. Thus, the import forms the condition for the formal object to intelligibly apply to the target in the given circumstances. Helm's notion of import is crucial for his understanding of affective intentionality as both responsive to and constitutive of import.

This is related to the claim that 'the very notion of direction of fit does not apply' (Helm 2011, 77n9). He calls this seeming contradiction in the intentionality of emotions 'the apparent paradox of simultaneous invention and discovery' (2000, 2), and it is on this basis that he arrived at his concept of emotions as felt evaluations. For Helm, emotions are felt evaluations insofar as they are commitments which both respond to and are motivated by import. The described mental state cannot be split into smaller parts and is at the same time perceptive and motivating. This notion brings the concept of direction of fit into question, as well as the categorisation of emotions into either cognitive or conative mental states:

> We cannot understand emotions merely as cognitive perceptions (with mind-to-world direction of fit) of an independently existing world; nor we can understand them simply as conative strivings (with world-to-mind direction of fit) projecting value onto the world [...] rather, the evaluative objects [...] are both that to which emotions are properly responsive and that to which we commit ourselves in such a way as to constitute that import. (2020, 228)

For Helm, the notion of direction of fit does not apply to emotions. Rather, affective intentionality essentially involves felt evaluations, which are unitary, undividable mental states, 'feelings that are simultaneously both responsive to and constitutive of that evaluation' (2011, 84). Furthermore, it is due to our commitments to import – described by him as a kind of cognitive background of personal value ascriptions/significance[30] – that, in the moment of emotional experience, we perceive something as either pleasant or painful.

Sabine Döring

Sabine Döring is also a proponent of affective intentionality.[31] She agrees with Goldie that the intentionality of emotions cannot be equated with that of beliefs or judgements, and similarly argues against the 'add-on' view proposed by composite theories: rather, the intentionality of emotions has to be seen as a 'synthesis' (Döring 2007, 376) of intentionality and phenomenality. However, she differs from Goldie's view in that she understands emotion by analogy with perception, as *affective perceptions*.[32] She does not equate them

30 Also described by Helm as the 'rational pattern of emotion' (2000, 10).

31 Cf. Döring 2003, 2007, 2009a, 2009b, 2010; Döring & Müller 2022.

32 Given that 'a perception also possesses both phenomenology and intentionality, and that its intentional content is also part of its conscious, subjective character' (2007, 376). She argues this with recourse to Peacocke (1992).

with perceptions,[33] but argues that 'in the rationalisation and justification of other states and actions the emotions play the role of perception' (376). In her view, emotions are intentional mental states because they have representational content, and because they are subject to correctness conditions. They are perceptions of the world as threatening, amusing, admirable, or sad; furthermore, our practical reasoning has to be 'suitably related to the intentional object' (377). How the intentionality of emotions functions in the rationalisation and justification of our actions is particularly important to her. And it is due to the correctness condition of an emotion that we can judge whether our, or someone else's, emotional response and issuing behaviour is appropriate. In experiencing indignation, for instance, a certain situation or behaviour is represented as unjust, which implies a respective evaluation of the intentional object. Correctness conditions related to the intentional object of a mental state also apply to beliefs and judgements. However, despite emotions similarly attributing some evaluative property to their intentional object, the attitude with which they attribute this property is different from that of beliefs or judgements, Döring argues.

She observes that emotions resemble perceptions and differ from beliefs in several ways. They share with perceptions that they do not stand in an inferential relation to the world, while beliefs and judgements do. According to her, emotions and perceptions share the same 'non-inferential logic': their experience is immediate rather than inferred. When we perceive a cat on the mat, we assume – under standard conditions, and unless we have reason to suspect otherwise – that there actually is a cat on the mat. We do not infer the cat to be there from the contents of our sense perception, but we non-inferentially take it to be there. According to Döring, emotions share this non-inferential role with perception. In an emotional experience we non-inferentially *perceive something as* threatening or amusing,

> there is no such thing as deductively inferring the experience of the behavior as amusing from the experience of its being excessively irritating [...] Instead, you immediately experience the behavior as amusing. (2009a, 243)

Furthermore, 'to perceive or to feel something as being a certain way does not imply regarding a certain proposition as true' (Döring 2009a, 244). While with

33 She differentiates emotional perceptions from sense perceptions based on a disanalogy noted by de Souza (1987, 150).

a belief, one regards something as true, in a perception or emotion it only appears to be true. There seems to be a cat on the mat, but in different light conditions it might turn out to be a rat. A snake appears to be dangerous, but it might turn out to be a harmless slow worm.[34] A thinker need not endorse the content of his perceptions or emotions; these things only seem to be the case. This is what she calls the 'default mode': that we take perceptions and emotions at face value. She explains this by attributing different formal objects[35] to beliefs and emotions: while 'truth' is the formal object of belief, it is not the formal object of emotion.[36] In her view, these mental states 'aim at truth' in a different way: 'to perceive or to feel something as being a certain way does not imply regarding a certain proposition as true' (2009a, 244). Rather, the formal object of emotion is different for every emotion type, for instance dangerousness is the formal object of fear, or injustice that of indignation.[37]

A further similarity she observes between emotion and perception is that neither need to be revised in the light of our better knowledge. She sees this as analogous to a difference between perceptions and beliefs, pointed out by Tim Crane: that perceptions need not be revised in the light of either other perceptions or beliefs that the perceiver has (1992, 151).[38] Crane illustrates this by perceptual illusions, such as the Müller-Lyer illusion. Here, we see two lines apparently of different lengths, although we know that they are the same length. Döring analyses cases of emotion in a similar way. The phenomenon is called the problem of 'conflict without contradition', and she discusses cases like Catherine Elgin's fear of dogs[39] and Hume's man in an iron cage[40] according to her understanding of affective intentionality. I will come back

34 Cf. Döring (2007, 378).

35 The formal object is typically understood as a property ascribed to the emotion's intentional object, which makes the intentional object intelligible as a token of a particular emotion type.

36 She argues this with recourse to de Sousa (1987, 122).

37 An idea that goes back to Kenny (1963, 189).

38 Perceptual illusions can illustrate this, and he is discussing the Müller-Lyer illusion of two lines that appear to have different lengths, despite having exactly the same measurements.

39 Cf. Elgin (1999, 146).

40 Hume considers a man 'being hung out from a high tower in a cage of iron (and) cannot forbear trembling, when he surveys the precipice below him, tho' he knows himself to be perfectly secure from falling, by the experience of the solidity of the iron, which supports him' (1739/40, 148).

to Döring's explanation of the problem of 'conflict without contradiction' in chapter 2.3.

In this sense Döring understands emotions as affective perceptions and argues against a propositional (what she calls 'judgementalist') understanding of intentionality in emotions. However, she shares with proponents of cognitive theories that intentionality necessarily entails an intentional object. The understanding of intentionality as 'having an intentional object' – what I call *objectual intentionality* – is different from that of 'having a propositional attitude', and I will specify this difference later. Döring's position clearly delineates the difference between a propositional and an objectual understanding of intentionality.

Jan Slaby

Another proponent of affective intentionality is Jan Slaby.[41] He assumes that the argument between cognitivist theories and feeling theories is settled, in as far as it is widely accepted that emotions are experiential states with intentional content.

> Besides being intentional, these affective states are phenomenologically salient and motivationally efficacious, and, moreover, hedonically valenced. (Slaby 2008a, 431)

With the term 'hedonically valenced', he refers to Bennett Helm, and he takes account of other theorists and their advances in this field, such as Sabine Döring's theory of *affective perceptions* or Peter Goldie's concept of *feeling towards*. However, in Slaby's view, they do not properly account for the bodily nature of the emotional experience. Following William James, he asks, 'What would be left of an emotion if all bodily phenomena would fall away?' For Slaby, the bodily feeling is the core of emotional experience, and he tries to capture the bodily nature of emotional experience without losing the central insight, that emotions are *intentional feelings*. That emotions are intentional feelingsmeans, in his understanding, that they are not directed only towards one's own body but also towards the world beyond. The bodily nature of emotional experience is what he calls the 'feeling body'. To achieve a more thorough inclusion of the feeling body in the understanding of emotions, Slaby lists five central features of intentional bodily feelings: diffuse localisation, world-directed

41 Cf. Slaby 2008a, 2008b, 2011, 2012.

intentionality, hedonic valence, self-consciousness, and motivational force. The incomplete list is supposed to show the complexity of bodily feelings and ask us

> to accept a class of bodily feelings which cannot be viewed merely as forms of awareness of clearly localized bodily processes but instead are crucially implicated in the awareness of something outside the body. (2008a, 434)

This fundamental bodily sensitivity towards the world, he claims, is an essential part of our experience. According to him, we have lost sight of the significance of the body through a tendency to take visual perception as a role model for our experience, and because of 'the narrowly mentalistic trend in Western thought' (ibid.).[42] For our concerns, I am going to focus on the aspect of world-directed intentionality.

Concerning the intentionality of emotions, Slaby first references Peter Goldie's concept of *feeling towards*. This concept describes how experiences of significant situations are echoed in the 'felt body', and how this exposes the subjective significance of the perceived in a specific way. Bodily feelings of this sort are seen as an inseparable aspect of the experiential state. Goldie first coined the concept of the unification view, which nowadays is a central feature of affective intentionality. He proposes not only that there is a contingent causal connection between perception and feeling, but also that they form a phenomenal unity: what Goldie calls a 'substantial event'. Slaby goes one step further in the understanding of the unification view to claim that 'the felt body is itself that through which we grasp what goes on around us' (2008a, 436). He alludes to Shaun Gallagher's concept of the 'body schema', which is contrasted with the 'body image'.[43] While the body image is the conscious image or perception we have of our body, the body schema is not something our experience is referring to; rather, it is *that through which* we are experiencing. It is understood as 'pre-noetic', by which Slaby means that

42 Slaby is talking about the 'felt quality' of our mental states as complementary to our 'visual bias'. He is referring to Hermann Schmitz, when he claims that the feeling body displays a diffuse but differentiable structure, in which one can distinguish different domains or regions. This vagueness of our inner experience he distinguishes from our external perception: we cannot 'feel-down' on our body in the same way we can 'look-down' on it. This point is discussed under the aspect of diffuse localization. (Terms in quotation marks are partly my translation.)

43 Cf. Gallagher (2005).

it is 'prestructuring conscious experience while being itself only indirectly accessible to introspection' (ibid.). As an example, Slaby describes the bodily experience of different emotions: when we feel offended, we have the physical impression of being 'oppressed' or 'pushed down' by an external force; and we feel an inner expansion when we feel joy, the feeling of occupying more space, which makes us at ease in our actual surroundings.Although we are aware of the abrupt narrowing of our body volume when we are afraid, and even if this implies more bodily changes, like a quickening heartbeat, our focus still remains on the object that evokes fear in us and is not directed to our body as such. This 'collateral' consciousness of our body is for Slaby one of the central aspects of the body schema. The bodily feeling is the essential structure of affective experience because it is the vehicle of perceiving significance: it discloses the world/ a situation/ an object to us and gives us access to it. And although (and precisely because) it is not in the focus of our experience, it adds indirectly to its content. The bodily feeling is essential to the way an event or situation is presented as significant to our experience, to its 'affective mode of presentation'. Furthermore, the feeling body is distinct from the felt body: the felt body is the perception of the body in pain, pleasure, itching, and so forth – it is the object of consciousness. The *feeling body* is not the object of consciousness, but rather the vehicle of the intentional feeling. It is disclosing the significance something has for us. Slaby describes it as that through which we become affectively aware of an object or situation. It is not an added-on judgement but is providing the structure of the affective process.[44]

Affective intentionality is understood by Slaby through the notion of intentional feeling: the directedness of our bodily feeling of the world, which discloses the significance of an object or situation to us. For him, affective intentionality is a different kind of relation to the world (*Weltbezug*) and the feeling body is the vehicle for this relation.

2.2.3 A different kind of intentionality?

Considering the different proponents of affective intentionality, two questions arose which I want to address: Do both claim that affective intentionality is a

44 He relates this to Merleau-Ponty's notion of the 'lived body', which is the structure-giving background of experience. This is not specifically referring to affective experience, but to experience in general, thus Slaby understands the feeling body as the 'hedonic' part of the lived body. Cf. Slaby (2008a, 441).

different kind of intentionality? And do both have the same 'general' understanding of intentionality?

Firstly, it has to be noted that the term 'sui generis' is not used as in 'one of a kind' but as in 'a different kind', and that all proponents of affective intentionality use it to point out a difference between emotions, and beliefs or judgements. However, I observed that some of the proponents claim that affective intentionality is an intentionality sui generis – that is, different from that of beliefs and judgements – whereas others claim rather that emotions are intentional states sui generis which must be analysed differently from beliefs and judgements. Both claims essentially pertain to intentionality, but the first holds that affective intentionality is a different kind of intentionality, while the second argues that emotions are a different kind of intentional state. The latter does not explicitly state that affective intentionality is a different kind of intentionality.

Bennett Helm claims that emotions are mental states sui generis. Although he is criticising the fact that in cognitive and composite theories 'the intentionality of emotions is [...] understood to be dependent on that of the constitutive beliefs and desires' (2001, 72), he is attributing this sui generis character not to intentionality itself but to the kind of evaluations that emotions, in his view, involve: namely felt evaluations. He explains that 'what makes something be a pleasure or pain is that it is an instance of a more general, sui generis kind which I have identified as felt evaluation' (2002, 32). In his view, affective intentionality is determined by 'how the characteris ation of its intentional content is crucial to understanding the phenomenology [of emotions]' (2020, 228). He does not in fact understand affective intentionality as a different kind of intentionality but rather explains how intentionality 'works' differently in emotions than in beliefs and desires. Helm introduces the concept of *import* in addition to several other kinds of intentional objects of emotion (like the target or the formal object of emotion), and it is the import something has for us which determines the focus of our attention. The intentionality of emotions works differently because, according to Helm, it is dependent on our commitments to import, which he also describes as the background 'rational pattern of emotion' (2000, 10). In effect, Helm does not claim that there is a different kind of intentionality of emotions, but that our emotions (identified by him as felt evaluations) are intentional mental states sui generis.

We find a similar view in the work of Sabine Döring, who also advocates a sui generis approach. She also attributes the term to emotions and not explicitly to intentionality: in her introduction to the anthology *Eine Philosophie der Gefühle* (2009) she describes emotions as 'mental states sui generis' (32), or 'perceptions sui generis' (51), and speaks of the 'sui-generis character of emotions' (37, all my translation). Furthermore, in a paper from the same year she is consistently speaking of emotions as 'mental states sui generis' (2009b, 241), and in the introduction to the special issue in which the paper appeared, of 'intentional mental states sui generis' (Döring and Reisenzein 2009, 201). In none of these works does she apply a sui generis character to intentionality. In fact, it is more accurate to say that she attributes it to the perceptions involved in emotion, which she refers to as 'affective perceptions'.[45] Döring argues against a propositional understanding of intentionality in emotions, but shares an important aspect of the understanding of intentionality with proponents of cognitive theories, namely that of 'having an intentional object'. Her position is exemplary of what I call *objectual intentionality*. An instance of objectual intentionality can be described as follows:

(oi) A mental state is intentional when it has an object it is 'about' or 'of'.

Objectual intentionality differs from propositional intentionality insofar as it does not conceive of intentionality as a propositional attitude, that is, as a relation to a proposition. A proposition is a statement that can be true or false, and a relation to such a statement consequently involves existential commitment to (or denial of) its objects. On these terms, the intentional objects of mental states are important for the rationalisation and justification of emotions *de dicto*, not *de re*. Objectual intentionality can be described as a mental state having representational content and being subject to correctness conditions; however, it does not necessarily involve an existential commitment to its objects. Thus, it does not reduce intentionality to mental states with a direction of fit; rather, its correctness conditions depend on its formal object or import. Döring does not claim that emotions possess an intentionality sui generis, but rather that they are mental states sui generis. To put it in her terms, emotions do not have 'truth' as their formal object but are subject to correctness conditions that differ from those of beliefs and desires.

On the other hand, Jan Slaby explicitly speaks of an intentionality sui generis: he claims that there 'is a sui generis category of human world-directedness' (2008a, 432), and, together with Achim Stephan and Henrik and Sven Walter, introduces this different kind of intentionality in their coedited book *Affektive Intentionalität* (2011) in the following way:

---

45 Cf. Döring (2009b, 51) andDöring (2007).

> It is a kind of relation to the world and oneself [Welt- und Selbstbezug] sui generis, which is so clearly distinct from the 'usual' kind of intentionality, as exhibited in unemotional propositional attitudes, like desires and beliefs, that it is appropriate to distinguish it also terminologically through the adjective 'affective'. (Slaby et al. 2011, 9, my translation)

Slaby understands the intentionality of emotions as a different kind of relation to the world: a 'world-directedness' that is established through what he calls the 'feeling body'. The feeling body is not the object of consciousness but rather the vehicle of the intentional feeling. It is through the feeling body that we are directed towards the world, and that we become affectively aware of an object or situation. According to Slaby, the intentional feeling is always 'part of' the world; through its embeddedness in the world, it is constituting our 'background-structure of experience' (Slaby et al. 2011, 31, my translation). These 'background feelings' are stable over longer periods of time, they mostly lack an object of emotion, and are often barely perceivable in terms of bodily changes. However, they are the result of our situatedness in the world and enable shorter, more intense, and clearly object-related emotions to form.[46] In this sense, affective intentionality is, for Slaby, an essentially different kind of intentionality. One characteristic of this kind of intentionality is what he calls *latentes Erschließen,* which literally translates to 'latent grasping':[47]

> Beyond the direct thematic grasping (as it is the case in active thinking, or the conscious focussing of our attention), there are indirect, 'unthematic' forms of experience. In the case of background-feelings, it is the framing of our experience and behaviour, which gives a particular qualitative note to the core of the consciously grasped [...] This latent grasping is not just the disposition to have certain experiences under particular circumstances, and feeling tendencies to act, it is rather the consciously present background-structure of experience. However, it is a non-reflexive form of experience, because it

46 Slaby bases his considerations on Sartre, who, in his view, is 'synthesizing Husserl's intentionality and Heidegger's Being-in-the-World' (2011, 26) to the idea of a situated understanding of existence which is questioning the mentalistic picture of an inside (person) and outside (world). Other recent positions in the philosophy of emotions, mentioned by Slaby, who reference the existentialist tradition, are Matthew Ratcliffe and Brian McLaughlin (cf. Slaby et al. 2011, 28n4).

47 In this passage, Slaby seems to use the words *erschließen* and *erfassen* interchangeably. In my translation, I am going to stick with the term 'grasping', since it is associated with notions of the body and touch, which are relevant to him.

> requires an active returning of our awareness, and the articulation of these latent structures, before we become explicitly aware of this affective framing of experience as such. (Slaby et al. 2011, 31, my translation)

Slaby is here contrasting the notion of 'latent grasping' as a nonreflexive aspect of our experience with cases of active thinking and focussed attention. The nonreflexivity, latency, and general situatedness of affective intentionality make it different from a propositional or objectual understandingof intentionality: it does not need an object, and it is permanent instead of momentary. Slaby does not explain how intentionality functions differently in emotions; rather, he proposes affective intentionality as a different relation we have to the world and oneself (*Welt- und Selbstbezug*).[48] With this different understanding of intentionality, he explains emotions as 'acute manifestations of significance' on a 'fundament of omnipresent latent grasping' (Slaby et al. 2011, 45).

There is a tendency in Slaby to associate with the views of Bennett Helm. He assumes that they share the same understanding of affective intentionality and thinks that Helm takes 'feelings as a sui generis category of world-relation' (Slaby 2008a, 433). On this point, I disagree: I think that there are differences between their understandings of intentionality. For Helm, intentionality works differently in emotions, but he does not speak of a different kind of intentionality. According to him, emotions are mental states sui generis which must be analysed on their own terms (namely, in the sense of felt evaluations). However, to call this a different 'category of world-relation' seems to be reading Helm's account too much in terms of Slaby's own understanding. For Helm, the object of emotion is a major concern in developing his theory: it is necessary in order to distinguish emotions from mere phenomenal states, and his notion of import constitutes the focus of our attention, which is introduced by him as a 'further object of emotion'.[49] Although both are proponents of the unification

48 Here, and in the quote on the preceding page, I chose to translate *Welt- und Selbstbezug* as 'relation to the world and to oneself'. This should not be understood as a relation which necessitates both relata to exist. I chose 'relation' over 'reference', because it emphasizes the situatedness in a context, and the duration of being-in-touch with something. This is related to the translation of Brentano's term *psychische Beziehung* as 'mental reference', which I will elaborate on later in my study.

49 In this view, emotions have several kinds of intentional objects. They are not simply responses to a given situation but are 'commitments to the import of both the target and, more importantly, the *focus*'. Helm further claims that 'the pattern of such emotions itself constitutes the import of that focus' (2011, 77).

view, Helm and Döring do not claim that affective intentionality is a different kind of intentionality. Rather, it became clear in my analysis that Helm and Döring share the understanding of 'objectual intentionality'.

There is a different implicit general understanding of intentionality at play in Helm, Döring, and Slaby. Although all are proponents of affective intentionality, their 'general' understanding of intentionality diverges. Helm and Döring distance themselves from a propositional understanding of intentionality, and given their emphasis on the object of emotion, their view is best described as 'objectual intentionality'. Slaby similarly considers a propositional model reductive; yet his notion of intentionality diverges from that of Helm and Döring. He characterises intentionality as a 'different relation to the world and oneself', which does not necessarily require an object and is therefore better described as 'directedness'.

However, there is another difference that can be observed in comparing these accounts: that of understanding intentionality as a property or a capacity. I think that Helm and Döring understand intentionality as a property of mental states, whereas Slaby understands it as a capacity of the mind. I will return to this distinction in chapter 4, where I further trace differences in the general understanding of intentionality by analysing the various paraphrases used for the concept.

2.3 Applying affective intentionality to Ref

Let us first return to why we considered affective intentionality in the first place. Emotional responses to fiction offer us two perspectives from which to consider intentionality: we can ask (A)' whether the intentionality of emotions is different from the intentionality of thoughts, and (B)' whether intentionality differs when directed at a fictional or factual state of affairs. Concerning the first question, I looked at the set of examples of Henry, Charles, and Aaron from Kendall Walton, and showed to what extent Walton's argumentation revolves around intentionality. It turned out that his understanding of intentionality is a major concern and constraint in his theory: it leads him to place us – as participants in games of make-believe – in the same realm as fictional characters, to maintain the intentional relation. Walton understands intentionality in terms of propositional intentionality, as the 'having of a propositional attitude'. This conception of intentionality also prevailed in (narrow) cognitive emotion theory at the time.[50] Due to this understanding of intentionality, Walton rejected the idea that Ref can be understood in terms of 'psychological attitudes', as this would imply a commitment to the existence of their intentional objects. Instead, he introduced the concept of quasi-emotions to account for Ref. I came to look at the concept of affective intentionality because it offers a different position: proponents of affective intentionality treat the intentionality of emotions as different from that of beliefs, judgements, and desires. Thus, it seems promising to apply affective intentionality to Walton's set of examples.

2.3.1 Unification view and inferential encapsulation

To apply the concept of affective intentionality to Walton's examples, we first have to formulate the conceptsinvolved more clearly. What do proponents of affective intentionality actually claim? As we have seen, there are considerable differences in their positions. Some argue that affective intentionality constitutes a different kind, while others maintain that intentionality simply functions differently in emotions. Helm and Döring understand intentionality as the 'having of an intentional object', which is not fundamentally different from propositional intentionality but rather subject to different correctness conditions. These differences prevent us from formulating a definition of affective intentionality. However, there is a common ground. Their respective claims are often formulated against cognitive and composite theories: for instance, they reject the idea that emotions are composed of different elements. There is no division between perception, judgement, and affect – these aspects are united in the emotional experience. The assumption that there is first a neutrally represented perception, which is then, in a second step, appraised and emotionally responded to, is precisely what leads them to call for a different kind of intentionality in emotions. Proponents of affective intentionality insist that phenomenality and intentionality are inseparably united in the emotional experience. The evaluation of a situation or object is not seen as distinct from what it is like to undergo that particular experience. Rather, emotions are understood as felt evaluations, affective perceptions, or intentional feelings, which unite intentionality and phenomenality. They evaluate a situation essentially *through experiencing it as* pleasurable or painful. This idea is called the

50 Narrow cognitivism in emotion theory claims that a belief or judgment is a necessary component for emotion, which typically entails an existential commitment.

unification view of content and quality, and it appears as the common ground for proponents of affective intentionality. The unification view can be formulated as follows:

(uv) The phenomenality of emotions and their intentionality are inseparable in the emotional experience.

The central characteristic of the unification view is that it does not treat intentionality as a separate element 'added on' to perception, as the idea of propositional attitudes suggests, but as something being *experienced as significant* in a certain respect. In the emotional experience our attention is 'gripped by the goodness or badness of something' (Helm 2002, 35): something is perceived *as* threatening, *as* amusing, *as* admirable, and so on. At least, this is the aspect that is mostly emphasized in arguing against composite theories. The unification view was further used to account for what Helm calls the *problem of emotionality*: to identify affective intentional states circle-free against nonaffective intentional states.[51] For both points, relevance attaches to what is 'included' as a result of this unity – that is, the evaluations, judgements, and beliefs inherent in the phenomenal perception of the situation.

There is, however, another aspect of the unification view. The inseparability of content and quality in the moment of emotional experience also leads to an 'encapsulation' of perceptual modes,[52] and thus to the exclusion of other information. This aspect is important because it can account for what Döring calls the 'problem of conflict without contradiction': that an emotion need not be revised in light of better judgment and knowledge. This logical possibility has to be accounted for when we consider emotions. Conflicts between different beliefs and conflicts between emotions and beliefs are both conflicts in content. However, in the latter, 'the subject need not revise (...) the mental states involved in the conflict in order to avoid contradiction' (Döring 2009a, 240–241). Furthermore, even in cases of revision, the emotion may persist. The sight of

51 Where should the specifically emotional aspect of an experience come from if its core – the intentional relation onto something in the world – is not in itself already emotional or affective? What could make an in itself affect-free belief or an emotionless judgement into an emotional state? All candidates, like bodily perceptions, motivations, or the intensity with which a conviction is taken to be true, already refer to emotions, and, thus, the explanation is circular. Helm claims that only if we take phenomenally and intentionality to be united in the emotional experience can we solve the *problem of emotionality*.

52 A term borrowed from de Souza (1987, 195).

a snake, for instance, typically provokes immediate fear because we perceive the snake as a threat, which makes us avoid sudden movements and cautiously keep our distance. Even after learning that a particular snake is not venomous, our fear might persist. We might no longer sweat and panic at its sight but still maintain a safe distance.[53] In this case, we know that the snake is not dangerous yet still perceive it as a threat.

> [It] would clearly be contradictory to judge that you are safe while at the same time judging that you are in danger (not safe). By contrast, it is not contradictory to judge that it is safe while at the same time feeling fear, that is experiencing the situation as dangerous. Although judgment and emotion are about the same thing and conflict with each other in how they represent that thing, the subject does not contradict himself, which allows the emotion to persist in the first place. (Döring 2009a, 241)

Döring continues to explain the difference between judgements/beliefs and emotions using Crane's distinction between judgements and perceptions, and examples of perceptual illusions.[54] Most relevant for our purpose is the difference she identifies in the formal object of emotions and of beliefs/judgements. While the formal object of beliefs is truth, emotions have different formal objects, depending on the type of emotion. The formal object of fear is danger, that of indignation is injustice, and so on. Formal objects are axiological properties that, in addition to individuating emotions, give them correctness conditions. In this way, we can explain why there is no contradiction between the content of belief and the content of emotion: they have different correctness conditions. While we know that we are safe looking at a snake behind thick glass, we might still be afraid of it and shudder at the sight – because it *looks* dangerous. This restricts inference between different kinds of mental states

53 We can recall Walton's example of Aaron and Greenspan's example of Frances here.

54 According to Crane, perception can be resistant to conclusive counterevidence because perception has a different kind of content. Its content is representational but non conceptual, and thus not subject to inferential constraints (if concepts are taken to be the inferentially relevant constituents of inferential states). This excludes inference but does not prevent it to function as evidence (Crane 1992, 147).

but does not prevent the exchange of information.[55] Some of this information might lead to revision, but not necessarily. Döring claims that emotions, like perceptions, have a non-inferential logic: their contents are non-inferentially given, and they only appear to be true (they are not yet endorsed). Furthermore, she claims that they both have a 'non-inferential structure' while 'the content of belief must be structured so as to relate beliefs inferentially to each other' (Döring 2007, 380). I believe that the latter claim can be further specified. Unlike perceptions, the different contents pertaining to the same emotion seem to be inferentially related, while remaining isolated from the rest of our beliefs and knowledge. The contents of emotion are gathered non-inferentially (through perception), but they have an internal inferential structure, similar to beliefs. This might not be obvious in the case of the snake, but it becomes clearer with more complex examples such as indignation or amusement. Döring refers to an example from Goldie (2005), explaining that if we find something amusing and our friend does not, we might explain to them what makes it amusing for us – but we cannot make them find it amusing. Amusement is not deductively inferred, she argues (Döring 2007, 382). However, we find something amusing because of a complex internal structure of beliefs, assumptions, and experiences, shaped by our language and culture. This internal structure is inferentially related. We find something amusing under certain conditions: there is no inference in the moment of amusement, but the structure that forms these conditions for amusement is inferentially related. This is why, to return to Goldie's example, we might find something annoying at one moment and amusing at another. The same sense content that we have non-inferentially gathered now applies to different correctness conditions. Furthermore, this inferential structure can alter over time: as we become more accustomed to a particular social or cultural setting, we, for instance, start to appreciate its sense of humour. We are not convinced by anyone that something is funny; rather, the internal inferential structure – and thus the correctness conditions of amusement – have changed. This can be described as the inferential encapsulation of emotions. Their contents are non-inferentially perceived and they have an inferential internal structure, yet they remain en-

55 Emotions have, among other capacities, the function of providing us with information: 'We actively relate ourselves to, and reflect upon, the information that we gain about our environment so as to treat the informational inputs as reasons for judgment and action' (Döring 2009a, 245).

capsulated from the inferential structures of other emotions and beliefs. This brings us back to the unification view.

The inseparability of content and quality explains not only the phenomenal unity in the moment of emotional experience (that is, what it includes), but also its encapsulation (that is, what it excludes). We can describe inferential encapsulation as

(ie) the inferential isolation of a mental state from other beliefs and knowledge that we hold.

This applies on different levels. There is a temporary encapsulation in the moment of emotional experience, and a more general encapsulation in the structure of different emotion types. The unification view can explain the former. To address the latter would require a discussion of the taxonomy of emotions, which cannot be attempted here. However, we could see that the unification view is closely related to inferential encapsulation: it constitutes its 'other side', so to speak. Together, I believe, they can account for cases of conflict without contradiction. And emotional responses to fiction are essentially such cases.

2.3.2 Explaining reduced behavioural output

Let us now consider whether the unification view would help us deal with Ref, namely whether it can explain the reduced behavioural output in Walton's set of examples. I will focus here on the cases of Charles and Aaron, since Henry's example is more related to concerns about the fictionality of works.[56] We can redescribe the scenes in the following way (added passages are in italics):

> Charles is watching a horror movie about a terrible green slime. *The camera is focussing on the oozing, slimy green mass. Charles perceives its material qualities as revolting, and its slow movement as unstoppable, swallowing everything in its path. He is fascinated by these strange qualities; they hold his attention. His eyes are fixed on the screen, and he finds it hard to look away. He is utterly focussed on the slime, and he cringes in his seat as it oozes slowly but relentlessly over the earth.* Soon, a greasy head emerges from the undulating mass, and two beady eyes fix on the camera. The slime, picking up speed, oozes on a new course straight towards the viewer. *The camera zooms in suddenly, and the surprising close-up is the final straw, pushing him over the edge. This moment tips the barrel, sending Charles*

56 I will come back to Henry's example when I discuss advantages of derived intentionality in chapter 3.

> *nerves into a frenzy:* he emits a shriek and clutches desperately at his chair. His muscles are tensed, his pulse quickens, his adrenaline flows. *He nearly jumps out of the chair. But just in that moment of impulse, when clutching the chair, he becomes aware that he is, in fact, in front of his TV and safe in his armchair.* Afterwards, still shaken, he confesses that he was 'terrified' of the slime. *His bodily sensations of disgust and fear might persist, but he did not follow the momentary impulse to jump up, or to run out of the room.*
>
> In one sense, Aaron realizes that airplanes are (relatively) safe. He says honestly that they are, and he can quote statistics to prove it. Yet he avoids travelling by air as much as possible. He is brilliant at devising excuses. If he must board a plane, he becomes nervous and upset. *Once, while already on board, the moment the doors were shut, he had a sudden feeling of panic and decided at the last minute to leave the plane. Although he knew all along that airplanes are safer than cars, he acted on his impulse to leave.* Aaron performs deliberate actions that one would expect of someone who thinks flying is dangerous, or at least he is strongly inclined to perform such actions. So, we legitimately infer that he does believe, at least on a 'gut' level, that flying is dangerous. *This contradicts his knowledge of crash statistics. However, he does not question his behaviour afterwards. He knows from experience that if he takes the plane, he will arrive completely stressed out: he would be confronted with thoughts about plane crashes and feelings of panic throughout the flight. Sitting in a car, by contrast, he feels safe – his body is not high up in the air, he can decide how fast to go, and he can steer the wheel himself. Despite his knowledge of crash statistics, he feels more in control of the situation.*

Let us briefly recall why Walton is considering these examples. His aim is to establish reduced behavioural output as a distinction criterion for Ref: he points out that Charles, confronted with a fictional situation, does not act, whereas Aaron, confronted with a factual situation – and 'despite his "intellectual" opinion to the contrary' – does act (Walton 1990, 198). As noted by Vendrell Ferran (2014), Moran (1994), and Werner (2020), reduced behavioural output cannot serve as a reliable distinction criterion for Ref, as Walton had claimed (1990, 31). Reduced behavioural output is important for Walton's argumentation, because it allows him to infer that quasi-beliefs (beliefs about fictional objects and situations) are involved in the emotional response. These quasi-beliefs can account for the intentionality of the emotion, but they motivate only quasi-emotions and quasi-actions, so to speak.[57] I have argued that a narrow cognitive theory of emotions, particularly in its understanding of propositional intentionality, imposes major constraints on Walton's view. As an alternative, I propose to apply the understanding of affective intentionality to the given scenarios. If we re-describe the two scenarios through the lens of the unification view, we can make sense of Charles's and Aaron's behaviour in a different way: it allows us to acknowledge that the intentionality of emotions 'works' differently from the intentionality of thoughts – and that this is something both cases have in common.

There are two main issues at stake: doxastic contradiction and practical irrationality. Both Charles and Aaron might be accused of holding contradictory beliefs. Charles seemingly believes and does not believe that the slime exists: he knows he is watching a science-fiction film and does not believe that there really is a green slime devouring a city. At the same time, his emotional response suggests that the slime is a threat – his muscles are tensed, his pulse quickens – and for the slime to be a threat, the narrow cognitive approach would claim he must believe it exists. Aaron, on the other hand, knows that flying is statistically safer than driving, and he truly believes it. Yet his fear of flying suggests a belief to the contrary. The shift from narrow to broad cognitivism in emotion theory[58] also shifted discussions around the paradox of fiction. Since the issue of doxastic contradiction in Ref is now largely resolved, the focus has shifted to practical irrationality (Vendrell Ferran 2014, 315). We looked at reduced behavioural output to discuss cases of practical irrationality. Both Charles and Aaron might seem to behave irrationally – Charles because his emotional response is triggered by something fictional, and Aaron because he acts against his better knowledge. Both scenarios can be understood as cases of 'conflict without contradiction', and I believe that the unification view, along with the related aspect of inferential encapsulation, can account for them.

Charles's behaviour appears irrational in two ways: his emotional response is to something fictional, and his behavioural output is reduced. These two aspects are intertwined. He perceives the slime as a threat and is spellbound

---

57 Walton does not use the term 'quasi-actions'; it is a paraphrase I have devised to fit his logic.

58 Namely, that a mental state which does not involve an existential commitment, like a thought or an imagination, can be the necessary cognitive component for emotion.

by its disgusting appearance and unstoppable movement.[59] His focus on the slime temporarily isolates him from the rest of his beliefs and knowledge. If the film suddenly cuts to a close-up,[60] Charles may shriek or clutch the arms of his chair, but he does not flee the room.

The unification view helps us make sense of this. First, it explains why Charles perceives the slime as a threat, despite it not being physically present. There is no neutral perception followed by evaluation. Instead, Charles perceives something *as threatening*: the content and quality of the slime are inseparable in his experience. The quality of oozing green sliminess provokes an immediate bodily reaction in him – nausea and revulsion. These bodily responses constitute a felt evaluation of the slime's perceived qualities: the bodily reflex to protect oneself from harm reflects an implicit assessment of those qualities as potentially dangerous. The feeling of revulsion indicates that something is threatening; and something is threatening because we feel repulsed by it. Yet this all takes place in front of a screen. Charles is not in physical proximity to the slime, and yet he still recoils in his chair and, at a sudden movement, nearly jumps up. This is where inferential encapsulation and Helm's notion of import become helpful. That something receives priority and is temporarily isolated from the rest of our beliefs and knowledge, can be described as inferential encapsulation. How this focus of our attention is established can, in turn, be explained with the notion of import. For Helm, import is at the same time constituted by and responsive to affective intentionality. The import – that is, one's caring about something – 'impresses itself on us by not only grabbing and holding our attention but also priming us to act appropriately' (Helm 2011, 80). It is the import something has for us that makes it a potential focus of attention. Thus, import forms the condition for the formal object intelligibly applying to the target in the given circumstances. It can explain how the slime – or rather, its qualities – is prioritised in Charles's perception over other elements of the film, over even the fact that it is a film, or the awareness of the room he is actually in. It accounts for Charles being utterly focussed, 'spellbound' by the slime: the felt evaluation of the slime's qualities

59 Why he is fascinated by, and to a certain extent perhaps enjoys watching, something threatening and disgusting is a different question, discussed under the notion 'paradox of horror'. I will go no further into this here.

60 Note the emphasis on the formal qualities of the film, like cut and image frame in this case, instead of using the sudden appearance of a murderer or similar. I am going to discuss the importance of formal qualities for Ref at the end of chapter 3.

is, for a moment, at the centre of his attention and temporarily isolated from the rest of his beliefs and knowledge. One could even say that this momentary isolation prevents his perception of the full situation around him, or at least a more neutral perception of it. This is the 'other side' of the unification view: while it *unites* certain aspects, it temporarily *excludes* others. If we understand Charles's fear of the slime as a case of inferential encapsulation, then his physical response is consistent with his perception of the slime's qualities, even though the slime is not physically present. Furthermore, even retrospectively, it makes sense that he would say, 'I was terrified of it', if we take the 'it' to stand for the revolting qualities of the slime and not the slime itself, because he was – due to the temporarily narrow focus of his view, brought on by emotion – indeed momentarily terrified of those qualities, even if he encountered them only on the screen. Here, the emphasis lies in the temporal priority of the felt evaluation of an object or situation before any more general evaluation.

This brings us to the second aspect of practical irrationality attributed to Charles: his reduced behavioural output. A close-up of the slime, focussing on its repellent qualities, makes Charles shriek and clutch his chair – yet it does not make him run out of the room. The unification view, along with the aspect of inferential encapsulation, helps us make sense of this reduced behaviour. Initially, Charles is completely immersed in the film, but he reorients himself when a disruption occurs: perhaps someone rustles their popcorn, or perhaps he becomes self-conscious of clutching the armrest in a moment of suspense. Through this disruption, his attention is no longer exclusively focussed on the film and the threatening qualities of the slime. He realises that he is safely and comfortably sitting in a cinema chair. His attention shifts, and the moment of 'encapsulation' ends. Although some of his feelings might persist, he would not act upon them fully. He may still feel disgust but will not run out of the theatre. This can be further clarified through Döring's notion of the 'default mode':

> we often operate in a default mode in which we take the content of our emotions at face value. Normally we rely on our perceptions because it does not occur to us to ask whether the conditions under which we perceive (such as the lighting conditions) are normal. [...] However, once we suspect that a sense perception or an emotion deceives us, we are ready to switch into a different mode. (2007, 379)

Charles, in this sense, is operating in default mode, taking the perception of the slime's qualities at face value. Due to inferential encapsulation, he is temporarily not concerned with the conditions under which he perceives these qualities. However, once this focus is disrupted, he becomes aware again of his actual surroundings and of the film; that is, of the conditions under which he has perceived the slime's qualities. He may still find these qualities disgusting, but he no longer perceives them as dangerous – at least not as dangerous to him, in his present situation. We can now understand why Charles does not act upon his initial impulse: the moment of encapsulation has passed, and he is (again) aware of the conditions under which he perceives the slime. As a result, the the same correctness conditions no longer apply to the content of his perception. The slime still looks disgusting, but it is no longer threatening. Of course, the idea that Charles might run out of the cinema is an exaggeration.[61] In reality, we are rarely completely immersed in fictional scenarios, but are well aware of the actual situation we are in. This discrepancy, experiencing things 'at a distance', is part of what makes emotional responses to art so interesting.[62] Here, at issue is the intentionality of Ref, and to what extent a different understanding of intentionality can explain reduced behavioural output. This is relevant because behaviour is often taken as a basis from which to infer intentional mental states. We could see that the unification view, and its aspect of inferential encapsulation, clears Charles of practical irrationality.

Similarly, one can argue, in Aaron's case, that he experiences being in a plane as a threat. Sitting in a heavy machine, with hundreds of other people, thousands of meters up in the air, easily counts as a threatening scenario. Perhaps he even saw a disastrous plane crash on the news or lost a loved one that way. When sitting in a plane, he is constantly beset by this feeling of threat, and by imagining other dangerous scenarios that could become the case. He can think of nothing else; he is utterly focussed on it. Since he must travel for work, he tries to gather objective information (statistics, etc.) and he comes to believe that traveling by plane is less dangerous than traveling by car. He is

61 Perhaps only when the Lumière brothers showed their first film of a train running into a station was it the case that people actually ran out of the cinema because of the projected scenes.

62 Volumes have been written on this aesthetic phenomenon, not only in philosophy, but also in film theory and psychology. And many exciting examples from contemporary culture could be discussed and compared. Aesthetic emotions pose very different problems and discourses than Ref. I discuss Walton's example, because it became prominent in the literature around the paradox of fiction. And I look at it only under the aspect of intentionality.

convinced that this is true, and he would state this belief to others. Still, he panics when he sits in a plane and would much rather drive to his appointments. He takes the car despite his better knowledge, and although the journey takes much longer. How can we save him from practical irrationality? Applying the unification view, we can argue that, in the moment Aaron sits in an airplane, his perception of the situation as threatening becomes the unwanted but absolute focus of his attention. The felt evaluation of being in danger is 'short-cutting the line': the experience of threat, and the focus on the threatening object, establishes an encapsulation of his thinking. His thoughts and feelings revolve around the dangerous airplane and thus exclude, or at least diminish, the reasonable belief that airplanes are safer than cars. The inferential encapsulation in the moment of emotional experience gives the thoughts of threat priority over other thoughts. What and why something is prioritised in these moments is partly a psychological question. *How* something receives priority, and how this is related to intentionality, however, can be explained with the notion of import. In Charles's case, it could explain how the qualities of the slime are temporarily prioritised, in his perception, over other things and their qualities in the film, or even over the room he is actually in. It can account for Charles being utterly focussed on the slime,and, similarly, for Aaron not being able to think about anything but all the possible airplane catastrophes. Applying the unification view to Aaron's scenario helps to explain the priority given to thoughts of threat over thoughts about statistics in the moment he is sitting in the airplane. And it becomes understandable that he eventually decides to travel by car, despite the knowledge that it is, in fact, statistically more dangerous. In the airplane, he is constantly confronted with thoughts of threat, whereas in a car, sitting at the steering wheel, he has the impression of relative control over the situation. The aspect of inferential encapsulation allows for Aaron to act upon his felt evaluation that flying is dangerous, while he actually believes that flying is less dangerous than traveling by car. Moreover, according to his experience, he will be completely agitated after a plane journey – relentlessly confronted with his panic – whereas he will arrive relaxed after a car ride. Prioritising such consideration over statistics of plane or car journeys cannot be called irrational either.

Although in Aaron's case we are not dealing with emotional responses to fiction, I considered it important to discuss his case in conjunction with Charles's. Walton constructs his argument of reduced behavioural output largely based on these two cases: Charles's fictional belief leads to reduced behaviour, while Aaron acts upon his irrational beliefs. Applying the unification view to both cases, we saw that, in fact, Charles and Aaron are not worlds apart. We can understand both their behaviours as cases of 'conflict without contradiction', which can be explained through the unification view and the related aspect of inferential encapsulation. How an object is temporarily isolated from the rest of our judgements and knowledge can be further explained by the notion of import. Thus, it is not the fictionality or factuality of the object of Charles's or Aaron's emotion that determines their different behaviour (reduced or not), but rather the particularities of how intentionality 'works' in emotions. Namely, emotions have different correctness conditions than beliefs or desires. And that emotional focus can create a temporary isolation from – and priority over – the rest of our beliefs and knowledge. The advantage of the unification view is that it allows us to explain how the 'reasoning' of our emotions, that is, the evaluations and judgements implicit in them, are sometimes prioritised and, for a moment, isolated from other beliefs and knowledge that we hold. It would be appropriate to call Charles's fear of the slime, and Aaron's fear of flying, subjective, since the evaluation of the slime, and that of sitting in an airplane, are based on subjective significance.[63] However, we cannot call them irrational – neither in the sense of doxastic contradiction nor in the sense of practical irrationality.

One doubt that could be raised against the unification view is that the insistence on not separating the phenomenality and intentionality of emotional experience may not pose a real problem. Just because something is considered separately in theory does not mean that one assumes these aspects to be actually separate. The separation of the object of emotion from an affective or evaluative attitude we have towards it can be merely for analytical purposes, and not as outrageous as proponents of the unification view sometimes suggest. Treating the intentionality and phenomenality of emotional responses separately in theory does not mean that the moment of emotional experience is actually split or compounded. We can continue to believe in its phenomenal unity. However,employing the unification view in the scenario of Ref showed that it is sometimes also useful to treat the different aspects of emotional experience as inseparable. It allowed us to explain how the intentionality of emotions can prioritise certain thoughts in the moment of affect, and how the unity of phenomenality and intentionality can create inferential encapsulation: a temporary isolation from other beliefs and judgements that we hold. Thus, we could make sense of the peculiar split second when we have the impulse to jump out

63 Helm also describes commitments to import as 'subjective significance'.

of the chair to rescue the heroine, or to run away from the slime. And we could also come to understand why it remains a split second, and why, ultimately, we do not follow the impulse.

2.4 Conclusion

The method of this study is to use Ref for analysing the concept of intentionality. Investigating intentionality in Ref in particular, and the role of intentionality in emotions in general, brings me to reconsider my initial question (A)': whether the intentionality of thought is different from the intentionality of emotion. Notably, the distinction between thoughts and emotions cannot be sharply drawn. Most of our mental states are intricate and intertwined conglomerates. The range of emotional responses is so wide – from fits and startles to cognitively complex emotions like jealousy or schadenfreude – that not only is the categorization of them difficult, but the definition of 'emotion' itself remains contested. How these different states are intertwined with cognitive elements is particular to each case, and it seems difficult to identify instances of 'pure' thought or 'sheer' emotion in our mental lives. Even if we succeed, these distinctions would only apply momentarily and in isolation. Our responses cannot be clearly divided into thoughts and emotions: they are interrelated and overlap in our experience. Furthermore, there is no *change* in our directedness whether we have a thought or an emotion about, say, Frankenstein's monster. We might entertain thoughts about the possibility of Frankenstein's monster in one moment, empathise with it in the next, and at the same time reflect on the formal aspects of how Mary Shelley evokes our empathy for the monster. In none of these cases does our relation towards the object seem to change – only how we relate to it.

Why, then, can it still make sense to look at the difference between thoughts and emotions in this respect? We began this chapter with the intuition that it makes a difference whether the objects of our emotions exist, or whether the objects of our thoughts do. This brought us to the example of reduced behavioural output in Ref, and to the question of the intentional object. I considered the concept of affective intentionality because, at first, it suggested that the intentionality of emotions is a different *kind* of intentionality. Applying the unification view (uv) and the related aspect of inferential encapsulation (ie) helped to explain the phenomenon of reduced behavioural output. However, since it turned out that some proponents of affective intentionality attribute the sui generis character to emotions rather than to intentionality, this does not help us answer the question (A)'. Both Döring and Helm adhere to the (uv) but understand intentionality as *objectual intentionality*, which only 'works' differently in emotions, while Slaby's understanding is better described as directedness, which does not necessarily need an object. In particular, Döring's considerations, regarding the problem of 'conflict without contradiction' in emotions, helped explain the reduced behavioural output of Ref. Since Döring does not understand affective intentionality as an intentionality sui generis, it is not necessary to assume a different kind of intentionality to account for Ref.

However, all proponents of affective intentionality emphasise the difference between intentionality in emotions and in beliefs. A belief or judgement requires an existential commitment towards its intentional object because of its truth condition. Otherwise, it would not be possible to assess whether the belief applies or not. In Döring's terms, this means taking 'truth' to be the formal object of beliefs. This does not apply to emotions which have different formal objects and, thus, different correctness conditions – such as danger for fear, or injustice for indignation. To have an emotion towards something does not necessarily entail that its object exists or the terms in which it is presented to us are true. In this sense, the intentionality of emotions differs from that of beliefs, since the latter involve an existential commitment. However, beliefs should not be equated with thoughts but are a particular kind of thought: thoughts with mind-to-world direction of fit. We can conclude that, based on the cases and theories discussed, the intentionality of emotions is not different from the intentionality of thoughts. Rather, beliefs and emotions do have different correctness conditions: beliefs necessarily imply an existential commitment, whereas emotions (and thoughts that are not beliefs) do not.

Furthermore, in comparing different proponents of affective intentionality, we saw that they have different implicit understandingsof intentionality. It seems that although there have been efforts to integrate the concept of intentionality into emotion theory, the 'general' understanding of intentionality has not been sufficiently questioned or differentiatedwithin it. I will return to this issue and discuss it further in chapter 4. First, however, I will turn to the other perspective of the paradox: question (B)', on whether there is a difference in intentionality when it is directed towards a fictional or factual state of affairs.

3. Dealing with fictions

> I had no intention of giving any of them a lift, naturally. However, as they wouldn't get off the road, I automatically moved my foot from the accelerator towards the brake. But then I thought, why? They weren't real. None of this was real. I wasn't really here, so they couldn't be either. It was absurd to treat them as real living people. So back went my foot on to the accelerator again.
> *Anna Kavan, Fog*

The concept of intentionality does not inherently have difficulty accommodating mental states that are directed to things that do not exist. That our mental states can be directed to both things that exist and do not exist is, rather, something the concept of intentionality must allow for and explain. 'In presentation, something is presented, in judgement something is affirmed or denied, in love loved, in hate hated, in desire desired, and so on', Brentano (1995, 68) famously states. The intentional object need not be true, nor does it need to exist. We can think of Pegasus, affirm that it is a winged horse (rather than a rabbit), and deny that it exists while desiring that it would. In such a case, our mental state is *about* Pegasus. It is for this reason that Brentano claimed that 'presentation' (*Vorstellung*) underlies all other mental states: whether something is affirmed, denied, or desired, our mental state is always presenting that thing, and thus can be said to be *about* it.In this sense, presentation is the fundamental form in which something is 'before our minds': every mental state involves the 'act of presentation',[1] which does not necessarily imply an existential commitment to what is presented. To this end, intentionality was described by Brentano as a particular kind of relation: a relation that does not require both relata to exist.

> If someone thinks of something, the one who is thinking must certainly exist, but the object of his thinking need not exist at all. In fact, if he is denying something, the existence of the object is precisely what is excluded whenever his denial is correct. So the only thing which is required by mental reference is the person thinking. (1995, 212)[2]

'Mental reference' is a translation of the term *psychische Beziehung*, which would literally translate as 'psychical relation', or, to avoid misunderstanding, 'mental relation'. Furthermore, it is precisely the aim of the cited passage – a part of Brentano's supplementary remarks from 1911 – to distinguish the German notions *Beziehung* and *Relation*, recognising that this might lead to confusion.[3] The 'mental relation' Brentano describes is not a truth-functional logical relation in which both relata need to exist. Rather, he emphasises that the one doing the thinking must exist, but what he is thinking about need not exist at all. Moreover, he dedicated an entire chapter to fictional objects, accommodating their nonexistence within the intentional relation in those supplementary remarks (Brentano 1995, 227–234).

These further explanations are presumably related to the fact that the intentional relation to nonexistent and fictional objects had become the subject of much debate among Brentano's pupils. Accounting for this relation was one of the reasons for Husserl (1901) to introduce the concept of *ideal objects*, and for Meinong (1910) to conceive his *Gegenstandstheorie*. The issue continued to preoccupy 20th-century philosophy to the extent that Tim Crane calls it 'the problem

1 'By presentation I do not mean that which is presented, but rather the act of presentation' (Brentano 1995, 60).

2 In the appendix for the 1911 republication of his work.

3 The German original quote is: 'Denkt einer etwas, so muß zwar das Denkende, keineswegs aber das Objekt seines Denkens existieren; ja, wenn er etwas leugnet, ist dies in allen Fällen, wo die Leugnung richtig ist, geradezu ausgeschlossen. So ist denn das Denkende das einzige Ding, welches die psychische Beziehung verlangt' Brentano (1925, 134). *Psychische Beziehung* has been translated into English as 'mental reference' in order to emphasise that it is not a typical relation. However, one is also losing something in using the term 'reference' here. I will elaborate on the difference between *Beziehung* and *Relation* in 4.2.3. and come back to the term in the last chapter as well.

of intentionality' (2001a, 22–23). Crane himself draws on the distinction between content and object initially made by Twardowski (1894) to address this problem: the object of a thought being that which one is thinking about, and the content of a thought being that which one is thinking. Some objects of our thought do not exist. Crane argues that 'while there is a sense in which one may be thinking, and yet thinking *about* nothing, there is no sense in which one may be thinking, and yet *thinking* nothing' (2001a, 33). Whether or not the object of a thought exists, it cannot be the case that a thought has no content – that its content is void. For Crane, this is a consequence of the fact that thoughts are relations to their contents. This is only to pick out one of the proposed solutions to the problem of thought about the nonexistent.

We can understand the paradox of fiction as one iteration of this problem, as it is the fictionality of the intentional object which constitutes one of the main issues for the debate. For Colin Radford, who, together with Michael Weston originated the modern debate around Ref in a paper presented to the Aristotelian Society in 1975, it is the nonexistence of Anna Karenina or Mercutio that, if we are moved by their fate, 'involves us in inconsistency and so incoherence' (Radford 1975, 78).This is also why some of the solutions to the paradox argue that our mental states are directed to something other than the fictional character: the surrogate-object-solution (Lamarque, 1981) holds that the *thought* of Mercutio is the intentional object; de Sousa (1987)claims that our emotion is directed at a *formal object* rather than the fictional character; and Choi (2001)argues that it is the *formal properties* of the film or book to which our emotion is directed. In this way, these theories try to avoid the problem posed by the nonexistence of the intentional object in Ref. But why is this considered a problem in the first place? We saw that, originally, intentionality was conceived in such a way that it could accommodate the fact that some of our mental states are directed at nonexisting objects. The question is not futile. I believe that the discussions around the paradox allow us to examine whether there is a difference for intentionality when it is directed towards a fictional or a factual state of affairs.

I have divided the paradox into two perspectives, both focussed on intentionality. In the last chapter I discussed (A)' – whether there is a difference between the intentionality of thought and the intentionality of emotion – and I now turn to (B)', the question of whether there is a difference in intentionality when it is directed towards a fictional or a factual state of affairs. In this chapter, I first revisit Colin Radford's (1975) *tragic-sister example* to investigate how the problem of nonexistence in Ref was initially formulated and to consider its possible implications for intentionality. Based on the results of this analysis, and endorsed by considerations by Frank Zipfel (2018), I emphasise the importance of the presentational aspect of fictions. We encounter fictions through artefacts.[4] It is through books, films, or performances that we learn about – that we 'get in touch' with – fictional characters and situations. Furthermore, the words and images in these works also appear to be 'about' something, as they represent certain characters and situations. This leads me to inquire whether we can speak of an intentionality of artefacts, and I examine the concept of derived intentionality to this end. Based on the intuition that words and other signs do not mean something by themselves but get their meaning from their maker or recipient, there is a tendency in philosophy to distinguish a primary/intrinsic intentionality of the mind from a secondary/derived intentionality of language, words, images, and so on. The concept of derived intentionality was introduced by John Searle in several essays between 1979 and 1983, which I first present and then discuss with respect to the advantages and disadvantages for explaining Ref in particular, and questions of intentionality in general.

3.1 Colin Radford's tragic-sister example

We relate differently to the account of something that actually happened than we do to the account of something made up – this is one of the basic intuitions which fuelled the theoretic interest in works of fiction, and which also led to the modern debate around the paradox of fiction.

> There is no problem about being moved by historical novels or plays, documentary films etc. For these works depict and forcibly remind us of the real plight and of the real sufferings of real people [...] What seems unintelligible is how we could have a similar reaction to the fate of Anna Karenina, the plight of Madame Bovary or the death of Mercutio. Yet we do. (Radford 1975, 69)

4 Artefacts can be objects like books or films, but can also have an immaterial form, as in performances or speech acts. I defined my use of the term in 1.4; however, I want to emphasise once more that, although most examples brought forward can be considered works of art, I am looking at them in their capacity of being fictions, and not in their capacity as artworks. I touched upon this in 1.4 and will comment further on the 'presentational aspect' of fictions in 3.3.2.

I want to go back to the start of the modern debate, namely, the papers given by Colin Radford and Michael Weston in 1975,[5] to examine how our involvement with fiction is entangled with issues of intentionality. What is referred to as Radford's *tragic-sister example* is of particular interest, since it is at this point that fiction enters his argument. By analysing this example, I aim to determine what implications this has for intentionality, and in what sense there may be a difference in intentionality depending on whether it is directed towards fictional or factual states of affairs.

Radford's text is structured in seven points, six possible solutions, and one conclusion (listed as point eight). He starts with observing men's concern for the fate of others as a brute fact (1). He then states that our being moved seems to involve a belief in the actuality, or likeliness of that fate, arguing that if the account of a terrible suffering we read about turns out to be false, we could not continue to grieve (2). He supports this claim with two further examples: someone telling us a tragic story about his sister, which subsequently turns out to be made up, and an actor friend giving a realistic performance of being in pain. In neither case can we be moved, he argues, because nobody is suffering, and we can at best applaud or criticise the performance (3). If, however, it turns out that the actor is reenacting the death agonies of a friend, this would legitimately horrify us. There is 'no problem' with being moved by historical novels or documentary films because they remind us of real events, Radford explains; whereas it seems 'unintelligible' for us to have a similar reaction to the fate of fictional characters, such as Anna Karenina – yet we do (4). He then considers if our Ref might be different from our response to real events, for example, the death of Mercutio from that of someone we know, due to the different ensuing behaviour: we might not be able to express our appreciation for the performance, while, at the same time, being moved to tears by it (5). However, Radford states that we 'shed real tears' for Mercutio, and we might be also ashamed of our reaction and appreciative of the performance at the same time. The problem, he reiterates, is that we feel pity for something that never happened (6). He asks how we can explain the 'seeming incongruity' of our being moved by Mercutio's death and Anna Karenina's fate (7). The bigger part of the text then offers the discussion of several solutions to the observed problem, which all remain unsuccessful, and Radford finally concludes that our 'being moved in

5 The two papers are listed in the Proceedings of the Aristotelian society under the same headline, and structured as Weston replying to Radford's paper, as was typical at that time.

certain ways by works of art, though very "natural" to us and therefore only too intelligible, involves us in inconsistency and so incoherence' (Radford 1975, 78). (8) This is how the problem which came to be known as the paradox of fiction was first articulated.

I won't give a summary of Radford's discussion of the different solutions.[6] Rather, I will focus on one aspect of how he sets up the problem, namely how fiction figures into it. In his argumentation, fiction is first mentioned in the tragic-sister example, and it is framed by two other scenarios:

> Suppose [...] that you read an account of the terrible sufferings of a group of people. If you are at all humane, you are unlikely to be unmoved by what you read. [...] But now suppose that you discover that the account is false. If the account had caused you to grieve, you could not continue to grieve. If as the account sank in, you were told and believed that it was false this would make tears impossible, unless they were tears of rage. If you learned later that the account was false, you would feel that in being moved to tears you had been fooled, duped. (Radford 1975, 68)

> Suppose that you have a drink with a man who proceeds to tell you a harrowing story about his sister and you are harrowed. After enjoying your reaction, he then tells you that he doesn't have a sister, that he has invented the story. In his case, unlike the previous one, we might say that the 'heroine' of the account is fictitious. Nonetheless, and again, once you have been told this you can no longer feel harrowed. Indeed, it is possible that you may be embarrassed by your reaction precisely because it so clearly indicates that you were taken in – and you may also feel embarrassed for the storyteller that he would behave in such a way. But the possibility of your being harrowed again seems to require that you believe that someone suffered. (Ibid.)

> Of course, if the man tells you in advance that he is going to tell you a story, you may reach for your hat, but you may stay and be moved. [...] An actor friend invites you to watch him simulate extreme pain, agony. He writhes about and moans. Knowing that he is only acting, could you be moved to tears? Surely not. Of course you may be embarrassed, and after some time you may even get faintly worried, 'Is he really acting, or is he really in pain? Is he off his head?'. But as long as you are convinced that he is only acting and is not really suffering, you cannot be moved by his suffering, and it seems

6 This discussion is interesting, however, since it mentions already most of the intuitions which were then developed by other theorists.

> unlikely as well as – as it were – unintelligible that you might be moved to tears by his portrayal of agony. It seems that you could only perhaps applaud it if it were realistic or convincing and criticize if it were not. But now suppose, horribly, that he acts or re-enacts the death agonies of a friend, or a Vietcong that he killed and tells you this. Then you might be horrified. (Ibid., 68–69)

In the first scenario, we read the account of terrible suffering, which turns out to be false. In the second scenario, someone tells us the tragic story of his sister, which turns out to be invented. In both cases, Radford states that if a seemingly factual account turns out to be a lie or invented, our emotion – our grieving or feeling harrowed – cannot continue. He points out that we might have other emotional reactions: we might be outraged or feel fooled in realising that we have been lied to, and we might be embarrassed at our initial reaction. However, these emotions are no longer *about* the terrible suffering in someone's tragic story. Rather, they are about the realisation that something we read or were told turned out to be not what we thought it was. The third scenario displays a similar structure to the first two, but inverted. Where the first two scenarios describe a change from factual to counterfactual/fictional, the 'acting example' describes a change from fictional to factual: an actor is simulating extreme pain, and we are sure that he is only acting, only for him to then disclose that he is actually reenacting the death agonies of someone he killed in war. While Radford states a decrease of empathy in the first two cases, there is an increase in the third. I want to point out two aspects, both related to how fiction is introduced to the problem: firstly, the comparison of a fictional to an untruthful scenario and, secondly, the particular structure of the three scenarios.

Fiction is for the first time mentioned in Radford's paper when he compares the tragic-sister example with the previous scenario, the reading of an account which turns out to be false: 'unlike (in) the previous one, we might say that the "heroine" of the account is fictitious. Nonetheless, and again, once you have been told this you can no longer feel harrowed' (1975, 68). He introduces the 'tragic sister' to compare this fictional scenario with the untruthful scenario in the first example. With this comparison, Radford wants to emphasise the nonexistence of fictional objects. It is a sensible comparison, since in both cases the object of our mental state does not exist. However, we have already seen in Walton that fiction cannot be equated with nonexistence. Furthermore, we are prone to overlook another fact when untruthful and fictional accounts are contrasted with factual accounts: that they are all 'made'. In all three cases, a state of affairs is accounted, not encountered. All the scenarios are mediated: the first one is read (it is not specified where), the second one is told by someone over a beer, and the third one is acted/reenacted. Focussing on the aspect of nonexistence – that is, the ontological status of the intentional object – we can lose sight of the fact that all three cases are mediated through some form of presentation. Furthermore, whether something is read in a newspaper or in a novel considerably affects the attitude we bring to the respective content. Often, it is not made clear to us whether something is fictional or factual; rather, we assume as much through the context and the form in which it is presented to us.[7] These considerations apply not only to 'materialised' forms, such as books and films, but also to immaterial forms of presentation, such as performances and speech acts in general. The account of a man telling us a tragic story about his sister is not merely an abstract content; rather, it was told in a certain manner and situation. Thus, whether we take it to be a fictional or factual account cannot only be due to the literal content of a story, but is equally determined by its form of presentation.[8]

The second aspect I want to point out in Radford's introduction of the problem concerns the particular structure of the three scenarios. In the first two examples, an account of seemingly real events turns out to be not what it seemed. In the first instance, we discover that the account we read is false; in the second, we are told that there is no sister, and the story was invented. In the third example, there is also a turn, but an inverted one: the simulation of pain turns out to be based on real events. The structure of the three scenarios can be summarized as follows:

(f-f) a seemingly factual account turns out to be false
(f-m) a seemingly factual account turns out to be made up
(m-r) a performance turns out to be based on real events

7 Authors sometimes intentionally attempt to confuse our sense of whether we are dealing with fiction or fact. Consider J. L. Borges adopting an academic style with footnotes, or even quoting from invented works, while the story itself is made up. In fact, this became an entire genre, including mockumentaries, for instance.That this playful confusion is possible only confirms that our common sense assumptions are informed by the style and form of a work's presentation.

8 Furthermore, *how* something is presented also plays a large part in whether we are moved by a book, film, or performance. However, this aspect of artistry is a vast topic and cannot be covered in the present study.

What Radford intends to demonstrate is that it makes a difference whether our emotion is directed at a factual or fictional state of affairs. He constructs the three scenarios in parallel, claiming that, in the moment we still believe in the factuality of events, we are emotionally engaged, whereas when the account turns out to be false or invented, we experience a decrease of empathy. The third scenario is designed to function like a cross-check: We might applaud the performance, but are not moved to tears because we assume that our friend is acting. When we learn, on the contrary, that he is reenacting real events, we are then horrified. Radford thus relates the factual to emotional engagement, and the fictional to the loss of it. From here stems the most questionable proposition of the paradox: that belief in the existence of an object is a necessary condition for emotion. This was already called into question (in the same paper) by Michael Weston, arguing that 'it doesn't follow that because some condition is necessary for our ascription of a certain feeling in a particular set of circumstances that it is also necessary for the ascription of that feeling in any circumstances whatever' (1975, 81). And he suggests that the opposite could be claimed on the same grounds: that belief in the existence of an object is, in fact, not a necessary condition for emotion because we are moved in those ways by fictional characters. The conclusions Radford draws, and particularly those which became proposition (2) of the paradox, were questioned from the start.

I already pointed out that we might overlook the mediateness of fiction if we only focus on the aspect of nonexistence in the tragic-sister example. Its inversion in the acting example has different flaws, but it leads to a similar conclusion. Firstly, I think that the third scenario is only seemingly an example of a fictional account because there is no narration given: our actor friend invites us to watch him 'simulate extreme pain, (he) writhes about and moans' (Radford 1975, 68). He does not tell us who is suffering, and he does not perform lines which would draw us into a story. He is performing pain; however, this cannot be called a fictional account. This would be an instance of confusing emotional responses to art with emotional responses to fiction. In the latter, a narration is required for something to qualify as fiction, whereas an artwork or performance may, but need not, involve elements that 'point beyond themselves'.[9] The simulation of pain is a performative act, but within the given context, it cannot be considered fictional. Still, our emotional involvement increases on learning that an abstract representation of pain is actually the reenactment of a real killing. This must be accounted for. We can see, however, that

9 I made this distinction in 1.4.

the symmetric construction of the three scenarios does not work out as Walton intended: the acting example does not qualify as Ref and thus does not function as a cross-check to the tragic-sister example.

Notably, I want to suggest that something else is responsible for the change in our emotional engagement. It does not hinge on the factuality or fictionality of the account but rather on the *turn* itself. The account of the terrible sufferings of a group of people – presumably an account of real events – *turns* out to be false. The tragic story about someone's sister *turns* out to be made up, fictional. In both cases, the status of the account changes. In the former, its logical status changes – that is, whether it is considered true or false – whereas in the latter, its discursive status changes – that is, whether it is considered factual or fictional. Radford argues that the decrease of emotion is due to the fictionality of events.[10] He states that 'if the account had caused you to grieve, you could not continue to grieve[, that] this would make tears impossible', and that 'the possibility of your being harrowed [...] seems to require that you believe that someone suffered' (1975, 68–69). However, in most cases of Ref, we are not deceived, but know from the outset that we are reading a fictional novel, or that we are watching a scripted film. In some cases, the impossibility of the setting – e.g., a science fiction film set at the centre of the earth or the surface of Mars – does not allow us, even for a moment, to forget the fictional nature of the narrative. Moreover, even if we initially believe we are watching a documentary and only later learn that the story was made up, we may still be moved when encountering it again. We can read certain books and watch certain films over and over again, and be moved each time anew, despite knowing that the story has been invented, and how it ends.

Radford does not specify where we are reading the first account, but since it is initially taken to be true, I assume that we are supposed to have encountered it in a newspaper or another context that typically conveys factual information. If we were to later discover that a newspaper article contains inaccurate information, we are unlikely to empathise with the sufferings of the people involved when we read it again.[11] There is a difference, however, if we learn that the ac-

10 He also suggests that we might get angry at the narrator, or that we might feel embarrassed because we have been fooled. This might occur, but it is not relevant to our question, since the emotions involved are not directed at the same object anymore, that is, the suffering group of people or the tragic sister.

11 In the next chapter, I will present a particular twist on this case as proposed by Frank Zipfel (2018).

count is fictional. Fiction is not committed to truth value. It is not false but made up: it is a pretended speech act with a different function. A work of fiction does not have the (primary) function of conveying information. Rather, it authorises us to imagine things 'for the sake of being imagined without being placed for the purpose of expressing a truth or referring to an item in the actual world' (Stecker 2010, 170). This marks a difference in the discursive status of the account.

I believe that the emotional response in Radford's examples is not due to the fictionality or factuality of the accounts; rather, it is due to the *turn* – to the change in the status of the accounts. In the first case, the truth value changes; in the second case, the discursive status of the account is changing. In both cases, we are taken aback and must reconsider how to relate to the new status of the accounted events. If the truth value of a statement changes, we can no longer relate to its content in the same way. If the discursive status of an account changes, we may at first be similarly taken aback. However, we might relate to it again: after realising that it performs a different function than we initially assumed, we may engage with the account, not for the purpose of gathering information, but, this time, for the 'sake of being imagined'. We may also be moved repeatedly by the same fictional story, because its function is not exhausted by having learned something new. Against Radford's claim, we can thus see that in the *tragic sister example* our empathy decreases, not because it is directed at fiction instead of fact, but because fact *turns* into fiction. The decrease of emotion cannot be attributed to the fictionality of the account per se, but rather to the change in the status of the account. We are taken aback and reconsider how to relate to the events described.

Let us consider the last scenario: while a decrease of empathy is attributed to the first two cases, an increase is attributed to the third. According to Radford, we are 'horrified' to learn that it is a reenactment of real events. This seems to contradict my suggestion that the decrease of emotion is due to the turn from one status to another. The third scenario also involves a turn, but in this case, Radford reports an increase in emotion. I have argued that the third example does not qualify as Ref and does not fit the same symmetric structure as the other two examples. However, we can think of other examples where this would be the case: for instance, the moment when we are informed about the real persons and events the film is based on, in the closing credits. Are we 'horrified' at learning that the film is based on real events? I think that we reconsider how to relate to the events and persons depicted in the film. Knowing that it is based on real events might make us feel more justified in some of the feelings we had, and other emotional responses might get readjusted in the light of this new framing. The initial reaction is, however, better described as 'taken aback' – that is, we literally take a step back: what we assumed turns out to be otherwise. It is not the fictionality or factuality of events per se, but rather the change in the logical or discursive status of an account – what I refer to as a 'turn' – that impacts our emotional involvement. The 'turn' takes us out of the relation we had to the respective account, and it is then seen within a different frame. We are confronted with it under these new conditions and, depending on the new status of the account – true or false, fictional or factual – we relate differently to it.

3.2. The presentational aspect of fictions

Both points I raised concerning Radford's examples emphasise the importance of what I want to call the *presentational aspect* of fictions: that we encounter fiction through a form of presentation. We are prone to overlook the mediatedness of all three accounts if we only focus on the ontological status of the intentional object. And the decrease and increase of emotion cannot be equated with the fictionality or factuality of events. Rather, it is the *turn* in the conditions of the accounts that makes us reconsider how we are to relate to them. These observations may be endorsed by Frank Zipfel (2018), who argues that having emotional responses to fiction implies that we are dealing with representations.[12] Following considerations by Moran (1994), Goldie (2003), and Matravers (2014), he claims that the aspect of representation has been neglected

12 The term 'representation' generally stands for artefacts, such as books and films, and he is speaking of the 'representational aspects' of an artefact to point out the way/means by which it is presented to us. For Zipfel, this also seems to include performances and speech acts, although these are not explicitly mentioned. In German, he is using the terms *Darstellung* and *Darstellungsaspekt*. I will refer to physical entities and performances with representational content as 'presentations' or 'artefacts' and their 'form of presentation' or 'presentational aspects' in order to avoid confusion with other uses of the term 'representation' in this paper (notably in Searle).

in the context of Ref.[13] The debate usually questions how our emotions towards nonexistent objects can be explained, while ignoring the fact that we encounter fictional objects only in the context of representations:

> The discussion is neglecting the fundamental and inevitable fact that we encounter the nonexistent only in representations and that representations always stand in a certain context of communication. (Zipfel 2018, 340, my translation)

He further points out that our emotional reactions to fictional or factual representations might be more similar than many theorists suggest.[14] Based on these considerations, he then reformulates the paradox of fiction as the paradox of mimesis: here, the emphasis lies in the difference between an emotion's being directed towards something 'confronted' or something 'represented', while the fictionality or factuality of the events is secondary. He also criticises what he sees as a tendency, in the discussion around the paradox, to take the ground for comparison as that between fictional and truthful factual accounts. This leads him to introduce a deceitfully factual account: the example of a newspaper article about Jimmy,a teenage drug addict which turns out to be made up by the journalist who wrote it.[15] Zipfel presented this text to several groups of students, asking them if they would classify it as a fictional or factual account. Besides voicing certain content-related aspects, in the end the groups always concluded that it must be a factual account, given the context in which it was presented (Zipfel 2018, 339). The context of the newspaper – an aspect of its presentation – figured as the main justification for believing in the factuality of the text's content. Zipfel concludes that the contextual framing of the artefact is crucial for Ref. He emphasises that our reception further depends on the medium, art form, and system of signs used,

13 He goes so far as to speak of 'emotions towards fictional representations' (*Emotionen gegenüber fiktionalen Darstellungen*) instead of 'emotions towards fictions' (cf. Zipfel 2018, 32711). Despite endorsing the aspect of presentation, I will continue to speak of Ref, since Zipfel's formulation can lead to confusion with emotions that have formal properties as their object.

14 Cf. Zipfel (2018, 340).

15 Janet L. Cook published the article title 'Jimmy's World' on September 28, 1980, in the *Washington Post*, and even received the Pulitzer Prize for this piece of journalism the following year. After the story was discovered to have been fabricated, Cook returned the prize.

and that these aspects must be studied more in order to adequately address the paradox of mimesis.

While most authors in the debate around the paradox focussed on the aspect of nonexistence in fiction, I want to instead focus, endorsed by Zipfel's point, on its *presentational aspect*. This aspect is not exclusive to fiction; rather, it is the condition of artefacts in general, whether fictional or factual. Content necessarily has a form of presentation: we encounter it in a book, a film, a speech act, or a painting. Furthermore, if it is written in a literary or a matter-of-fact style, and if it appears in a paperback or newspaper, even certain conventions of typeface and production may indicate whether we are dealing with a fictional or a factual account. In Jimmy's case we saw that readers typically cooperate with the form of presentation: we do not begin researching each case we read about in an article, but generally take the text to be a factual account because it is published in a newspaper.[16] The form of presentation concerns the context, medium, style, and aspect under which content is presented. Furthermore, the form of presentation can be indicative of the logical and discursive properties of an account – that is, if we take the account to be true or false, fictional or factual. Since it is necessarily through a form of presentation that we have access to a content, a content cannot be sufficiently individuated without its form of presentation.

These aspects are crucial for our relation to artefacts, and I will discuss them in more detail later (particularly in chapter 3.3.2). First, however, I want to make a more general point: the presentational aspect concerns not only *how* something is presented, but *that* something is presented to us. It is only through some form of presentation that we have access to content. The fact that we encounter fictions through a form of presentation is crucial for considering perspective (B)', because it influences how intentionality must be analysed in this context. Since we encounter fictional (as well as factual objects) through a form of presentation, we must take this mediatedness into account. What does it mean for intentionality to be directed to a content *through* something else? And is an artefact – be it a book, film, or performance – itself something that is 'directed'? Artefacts are representational,[17] and seem to exhibit a ca-

16 However, we should perhaps be less 'cooperative', given the recent increase in fake news.

17 There are of course also artefacts which are not representational; however, I defined the way I use the term 'artefact' as 'physical item or event with representational content'. Cf. chapter 1.4.

pacity similar to that of mental states: they are 'about' things. Fiction-directed intentionality is usually treated as the directedness of a mental state towards a fictional state of affairs, the issue being the ontological status of the intentional object. In the discourse around the paradox, the presentational aspect of the intentional object has been neglected; it was not taken sufficiently into account that fictional content is mediated through a form of presentation.[18] This opens a range of further issues. One fundamental question that must be considered in this context is whether artefacts themselves exhibit a form of intentionality – whether they are themselves 'directed'. Does it make sense to attribute intentionality to artefacts as well, or to analyse them in these terms?

Before we tackle this question, however, we must clarify whether the presentational aspect is a necessary condition when dealing with fiction. As observers, we encounter fiction through artefacts like books, films, and performances, and this condition is not restricted to fictional content: it applies also to factual events we learn about in newspapers, documentaries, and textbooks. But we also encounter factual events through direct experience. Can we exclude instances of an unmediated encounter with fictional characters or events? If we set aside for the moment the point of view of the observer, which is usually considered in conjunction with the paradox, and consider instead the situation of the maker, this question does not seem so far-fetched. What is the relation between an author and a fictional character, and could there be instances where this encounter occurs 'directly', not mediated through some form of presentation? Can we think of an instance of unmediated fiction-directed intentionality?

If we consider the process of creating a character, there might be instances in which the author thinks about a character without mediation. Mary Shelley describes in the introduction to the third edition of *Frankenstein* (1831) that she was looking for an idea for a ghost story because of a contest between Lord Byron, Percy Shelley, Dr. Polidori, and herself. She recounts how, after a late-night discussion about Darwin's experiments and the phenomenon of Galvanism, she could not find rest:

> I did not sleep, nor could I be said to think. My imagination, unbidden, possessed and guided me, gifting the successive images that arose in my mind with a vividness far beyond the usual bounds of reverie. I saw – with shut eyes, but acute mental vision, – I saw the pale student of unhallowed arts

18 An important exception is Zipfel's (2018) account.

> kneeling beside the thing he had put together. I saw the hideous phantasm of a man stretched out, and then, on the working of some powerful machine, show signs of life, and stir with an uneasy, half vital motion. (Shelley 1818/2018, 241)

She further describes how she opened her eyes in terror, wanting to replace the image with the reality around her, but the image persisted. She could still see the scene of her imagination, and, returning to the ghost story contest, she thought, 'Oh! if I could only contrive one which would frighten my reader as I myself had been frightened!' (ibid.). So she decided to make this 'spectre' her story. Could one say that her being terrified of that 'phantom'[19] is an instance of unmediated fiction-directed intentionality? The scene Mary Shelley imagined that night may not count as fiction in the sense of a work of fiction, as it was not yet written down, but it can be considered as 'imaginary'. However, this is only the beginning of the story. From that terrifying vision, she went on to develop the characters of Victor Frankenstein and his monster. The complexity of the latter – the rejected and abandoned creation – suggests that Shelley started to engage with the monster. The majority of the story is written from the monster's perspective and describes his experiences and feelings in a compassionate way. What at first terrified her was developed into something readers can empathise with and pity. This suggests that she came to empathise with the monster herself in the process of writing.

Octavia Butler, a more recent science-fiction author, addresses something similar in an interview: somewhat confessionally she admits that she did not like one of her main characters, Lauren Olomina, much in the beginning. She could not relate to her at first. But in the end, Butler says, 'I got to the point where I liked my character far too much'.[20] 'In the end' probably means when she had finished writing the book. This, similarly to Shelley's account, suggests that she developed empathy for her character in the process of writing. Only someone who has never been in the position of writing a text, of creating a story, would imagine the author developing a character and an entire storyline for it in their imagination, then writing it down in one go. Rather, such

19 She is using both 'phantom' and 'spectre' in her description, but it is not clear whether she is referring to the whole scene or specifically to the monster.

20 From an archived recording integrated in the podcast *Octavia Butler: Visionary Fiction* (2021, min 57:16 and min 58:51).

a process typically involves rereading already-written parts, as well as changing, deleting, rewriting, and recomposing them. This means that the creation of a character or narration itself involves mediation. The mental states of the author are not directed towards something fictional without being mediated through a form of presentation, because the fictional is made-up, and the process of making up a character or story involves the repeated rereading of the fictional content in provisional forms of presentation.[21] This implies that not only the reader but also the author encounters fictional characters in a form of presentation. The initial speculation that the author might be 'in direct touch' with a fictional object can thus be set aside. We can conclude that we necessarily encounter fictional characters or events as mediated through presentations.

What about Mary Shelley's spectre, the vision of Frankenstein creating a monster? The initial apparition seems to share its nonexistence with fiction. However, we must recall that the fictional cannot be equated with the nonexistent: while both may be called imaginary, the latter concerns truth-value, whereas fictional discourse has a different function. It is an invitation to imagine things 'for the sake of being imagined' (Stecker 2010, 170). In this sense, Shelley's vision does not count as a fictional object and thus does not fall under the phenomenon of Ref discussed in this study. Only later, when Frankenstein's monster is described and 'created' by her throughout the pages of the novel, does it become a fictional object. However, if we consider the phenomenon of her having that vision, we may assess that her fear of such a vision – of something illusory and nonexistent – is indeed unjustified and incoherent. Our emotional responses to fictions are a different case, and must be assessed differently. Thus, Mary Shelley's spectre is not a counterexample. We may therefore conclude that one major point concerning Ref is that, when we deal with fictions, we necessarily deal with a form of presentation.

The presentational aspect of fiction has been neglected in the discussions around the paradox. However, I believe that it is crucial for our understanding of Ref. This shifts the emphasis in perspective(B)'from the issue of nonexistence to that of mediatedness. What implications and consequences does it have for intentionality if we take the presentational aspect in Ref more seriously? Should we attribute intentional properties also to artefacts? And what

21 We can recall Walton's definition of the fictional here. For him, the fictional serves as a prop in games of make-believe: it is made for, and we could now also say through, our participations in games of make-believe.

would it mean for the intentionality of our mental state if it is directed *through* something that is itself directed? We can reformulate (B)' accordingly:

(B)" Is there a difference in intentionality depending on whether it is mediated or immediately directed at a state of affairs?

3.3 The intentionality of artefacts

To take the presentational aspect of a (fictional or factual) account more seriously, we must inquire about the intentionality of artefacts. It is necessarily through forms of presentation that we have access to certain content; it is, for instance, the only way we can learn about fictional characters and situations. Artefacts seem to display a similar capacity as our mental states: they are 'about' things. But do artefacts themselves exhibit a form of intentionality, or does it even makes sense to analyse them in these terms? Intentionality is typically only attributed to mental states, not to physical items or events. Would it be helpful to attribute intentional properties to artefacts, and in what way could we understand such properties?

Words and images are also 'about' something; they are pointing at or are directed towards something beyond themselves. It seems to be a common intuition that a thought about Frankenstein's monster and a sentence or picture of Frankenstein's monster have something in common, namely, that they are both 'about' Frankenstein's monster. This suggests that we can attribute the notion of intentionality to both mental states and artefacts. However, there are differences in the way the two – mental states and artefacts – are directed.To examine this difference, let us consider something Aristotle says on the first pages of his *De Interpretatione*:

> Spoken words are the symbols of mental experience and written words are the symbols of spoken words. Just as all men have not the same writing, so all men have not the same speech sounds, but the mental experiences, which these directly symbolize, are the same for all, as also are those things of which our experiences are the images. (Aristotle 1928, 16a)

We can find two claims concerning the difference between writing/speech sounds and mental experiences in this passage: while our writing/speech sounds differ, our mental experiences are the same. Furthermore, writing/speech sounds symbolise our mental experiences, whereas our mental experiences are the 'images of things'. The second point is particularly interesting for our concerns, as it suggests that mental states *relate* to things differently than artefacts. The difference, this passage suggests, is that the intentionality of artefacts is derived from the intentionality of our mental states, while our mental states are 'directly'about things, in the sense that they are the images of those things. The latter relates to Aristotle's theory of sense-perception, according to which the mind takes on the form of the perceived object.Not concerning itself with exactly how mental states are 'about' things, the passage implies that the intentionality of artefacts is mediated through the intentionality of the mental state: writing/speech is directed towards things by symbolising a respective mental state. This resonates with considerations about the aspect of presentation in the last chapter. The idea that a sentence, for instance, is an expression of a thought, and that its intentionality is somehow derived from the intentionality of a respective state of mind, seems quite intuitive. We can easily imagine an author thus producing speech sounds or writing. But what happens in the case of thereader? Being confronted with a sentence, we might be able to say that it is 'about' something, but where does the sentence derive its intentionality from? Does it lie in the written words, and do we somehow recognise it? Do we assume that an author 'put' intentionality into those words, or do we ourselves 'put' intentionality into the words when reading them?

These questions touch on vast domains in language philosophy, such as meaning and reference. Considering them here can, of course, at most – and hopefully – merely add an aspect to these discussions. Furthermore, the concept of intentionality is traditionally attributed to mental states: it was reintroduced by Brentano to delineate the realm of the mental, claiming that it is the 'mark of the mental'. Still, given the common property of mental states and artefacts to be 'about' something, it might turn out to be useful to also analyse artefacts in these terms. One particular and elaborate theory, which applies the concept of intentionality to speech acts (and other artefacts), was proposed by John Searle: namely, the concept of *derived intentionality*. It is meant to explain the nature of speech acts, and to account for the relation between the intentionality of speech acts and the intentionality of the mind. The general idea is that there is a primary/intrinsic intentionality of the mind and a secondary/derived intentionality of language, words, images, and so forth. On this view, cultural artefacts like films, books, images, and sentences do not mean something by themselves, but come to mean something through the intentionality of the mind. As I concluded in the last chapter, we must account for the presentational aspect of fictions, that is, that we encounter them through a form of presentation. This led to the question of whether we can apply the concept of intentionality also to artefacts. The concept of derived intentionality proposes such an application. As such, it can possibly help us ascertain how to approach the the implications and consequences of a mental state being directed through something that is itself directed.

I will first introduce Searle's concept, then consider whether it could be useful for explaining Ref in particular, and if it could help address the presentational aspect of artefacts in general.

3.3.1 John Searle's concept of derived intentionality

Based on the intuition that words and other signs do not mean something by themselves but get their meaning from their maker or recipient, there is a tendency in philosophy to discern a primary/intrinsic/original intentionality of the mind from a secondary/derived intentionality of language, words, images, and so on. The concept of derived intentionality was introduced by John Searle in several papers published between 1979 and 1982, and in his book *Intentionality: An Essay in the Philosophy of Mind* (1983). It has been prominently discussed in the context of artificial intelligence and the computational theory of mind. Nowadays, it is still sometimes used as a distinction in the context of consciousness and phenomenal intentionality, for example, alongside the claim that there are instances of underived intentionality, or underived content. Searle introduces the difference between intrinsic and derived intentionality as follows:

> Speech acts have a derived form of Intentionality and thus represent in a different manner from Intentional states, which have an intrinsic form of Intentionality. (Searle 1983, 5)[22]

> The actual performance in which the speech act is made will involve the production (or use or presentation) of some physical entity, such as noises made through the mouth or marks on paper. Beliefs, fears, hopes, and desires on the other hand are intrinsically Intentional [...] speech acts have a physical level of realization, qua speech acts, that is not intrinsically Intentional. (27)

22 Searle is capitalizing 'Intentionality' and 'Intentional' to distinguish the meaning of directedness from the meaning of intending to do something. I will keep his capitalization in the quotes but continue to refer to intentionality myself in lowercase letters.

> An utterance can have Intentionality, just as a belief has Intentionality, but whereas the Intentionality of the belief is intrinsic the Intentionality of the utterance is derived. (27)

Phenomenal range

Firstly, it is necessary to clarify the phenomenal range of things to which Searle applies the term 'derived intentionality'. Although he is mostly referring to speech acts – ultimately his theory of intentionality is meant to ground his speech act theory – his account of derived intentionality is not limited to them. Other physical phenomena can have derived intentionality, such as marks made on a paper. At another point he extends the list to 'maps, diagrams, laundry lists, pictures, and a host of other things' (Searle 1983, 26). In fact, one of the papers dealing with intentionality, and developing the concept of derived intentionality, which leads up to the book published in 1983, is an essay on pictorial representation: in the spring 1980 issue of *Critical Inquiry*, which focussed on the language of images, Searle published a paper called '*Las Meninas* and the Paradoxes of Pictorial Representation'. Applying the methods of philosophy of language to Velázquez's painting, he shows that the painting is self-referential: the question 'what is the painting about?' must be answered with 'it is about itself'. Here, he addresses the issue of derived intentionality as follows:

> [...] utterances, writings, and pictures are not in that way intrinsically Intentional (as beliefs, desires etc. are); they are physical phenomena in the world like any other physical phenomena. And the central problem of the philosophy of language is to explain how the physical can become Intentional, how the mind can impose Intentionality on objects that are not Intentional to start with, how, in short, mere things can represent. (1980a, 250–251)

He assumes that a lot of things are capable of having, or obtaining, derived intentionality, and he is interested in how intentionality can be imposed by an agent onto something – how that something can be 'made' intentional. He considers intentionality a 'ground floor property of the mind', and that it is impossible to give necessary and sufficient conditions of intentionality in terms of simpler notions because we cannot find a 'neutral standpoint' (1983, 26) from which to observe the relation between our mental states and the world. However, we must not replace 'speech act' too rashly with this list of other things. His strategy is to use our understanding of how speech acts work to explain how the intentionality of the mind works. That is, he looks at intentionality through the perspective of speech acts. Since we are interested in a general notion of derived intentionality which can also be applied to images and other forms of presentation, it is important to keep this point in mind. There are instances in which Searle applies the term 'derived intentionality' to a wider range of phenomena; however, his analysis of the term is conducted through the perspective of speech acts, and linguistic meaning. I will call this a *perspective restraint*, and we must consider this restraint before (potentially) taking on his claims for a wider purpose.

Understanding and use

After having looked at the range of phenomena to which the term 'derived intentionality' is applied, we can ask how it is defined. Searle does not provide a proper definition at any point, so I will trace his understanding by examining how he uses the term and why he introduces this notion. We can divide the following considerations into aspects that intrinsic and derived intentionality share, and aspects in which they differ.

The general characteristic that, according to Searle, intrinsic and derived intentionality have in common is that they are forms of representation. In one paper, 'representing' is part of the list of paraphrases included when introducing intentionality: 'the feature which certain mental states and events have of being [...] *directed at*, or *about*, or *of*, or *representing* certain other entities and states of affairs' (1984, 87). This does not mean that intentionality can be reduced to the notion of representation, but the latter seems to be, for Searle, the main feature intentional mental states and speech acts share. Regarding his understanding of intrinsic and derived intentionality as forms of representation, he makes the following claims:

> Intentional states represent objects and states of affairs in the same sense of 'represent' that speech acts represent objects and states of affairs. (Searle 1983, 4)

> Intentionality is that property of the mind (brain) by which it is able to represent other things. (24)

> [...] the way that language represents the world is an extension and realization of the way the mind represents the world. (166)

> The main function which language derives from intentionality is, obviously, its capacity to represent. (175)

Further similarities between speech acts and intentional states are: (1) the distinction between propositional content and illocutionary force (for mental states he calls this the representative content and the psychological mode or manner in which one has that content); (2) that both have direction of fit (although there are also 'null cases', e.g., sheer expressions, and 'double cases', e.g., declarations); (3) that both have conditions of satisfaction (although they don't always have to be satisfied, e.g., in lies, mistakes, and fictional discourse); and (4) that the intentional content of the mental state figures as the sincerity condition of the speech act (i.e., the performance of the speech act necessarily expresses the corresponding intentional state). He goes on to define this set of similarities as a constellation of logical notions for which the term 'representation' serves as shorthand. Thus, the term 'representation' is used by him in a very specific way: it is neither understood in terms of depiction, nor as an artefact, nor as an account of meaning. Rather, it stands for this set of logical notions he identifies. Searle's use of 'representation' can be summarised as something that has a content and a mode: the content determines the conditions of satisfaction, and the mode determines its direction of fit. Furthermore, speech acts are representations in the sense that their representational capacities are derived from the representational capacities of mental states.

To understand how Searle arrives at the notion of *derived* intentionality, we must further examine how intentional states and speech acts differ. This disanalogy is closely linked to the problem of meaning. He observes that meaning applies to sentences and speech acts, but not in the same sense to intentional mental states: 'John believes that p' can stand on its own, whereas 'John means that p' requires completion in the form 'by saying such and such' (1983, 28–29). It makes no sense to ask what a belief or desire means, because meaning exists only if there is a distinction between intentional content and the form of its realisation. Furthermore, speech acts differ from mental states because they exhibit what Searle calls a 'double level of intentionality':

> There is first of all the Intentional state expressed, but then secondly there is the intention, in the ordinary and not technical sense of that word, with which the utterance is made. (1983, 28)

To account for this double level of intentionality, he references Grice's theory – that for a speaker to mean something by an utterance is for him to have a set of intentions directed at an actual or possible audience.[23] However, Searle rejects the idea that the relevant intentions are those aimed at producing certain effects in the audience.[24] He disagrees because he believes it is crucial not to confuse representation with communication, and also not to conflate the intention to make a statement with the intention to make a true statement. Any account must allow for the possibility of lying, of making a statement while failing to make a true statement, or of making a statement without caring whether the audience believes or understands it (1983, 168). Searle considers the difference between saying something and meaning it and saying something without meaning it. He illustrates this with the example of saying 'es regnet' to practice German pronunciation or saying it because it is raining (or because it is relevant whether it is raining or not). The difference is that in the first case, the utterance has no conditions of satisfaction. He concludes that the essence of statement-making is *representing something as being the case* – not communicating one's representations to an audience. The intention to communicate is just one among various possible intentions, like promises, orders, or expressions. According to Searle, we must consider the difference between an act and a state: acts differ from states because they involve the production, presentation, or use of some physical entity. And the action of producing, presenting, or using such a physical entity implies an intention on the part of the agent.

> It is the performance of a speech act with a certain set of intentions that converts the utterance act into an illocutionary act and thus imposes Intentionality on the utterance. (Searle 1983, 29)

Searle understands intentionality as the 'directedness' or 'aboutness' of mental states. However, his view differs from Brentano's in several respects: (1) For him, only some, not all, mental states and events have intentionality. To distinguish mental states with intentionality from those without it, we must be able to answer questions such as 'What is the mental state about?', 'What is the mental state of?', 'What is it a mental state that?' (2) He distinguishes intentionality from consciousness, arguing that many conscious states are not intentional,

---

23 Searle is referring to Grice (1957); note that 'intention' here is meant in the ordinary sense of the term.

24 Cf. Searle (1983, 161).

and many intentional states are not conscious. The confusion arises from the use of the word 'of': saying that we have an experience of anxiety uses 'of' differently than in saying that we are afraid of snakes. In the latter, the state is not identical with its object. For Searle, it is characteristic of intentional states that there is a distinction between the state and what it is directed at, about, or of.[25] (3) It is important to note that for Searle the intentionality of language is derived from the intentionality of the mind, implying that intentionality is not essentially and necessarily linguistic.[26] Although he locates intentionality in the mental, his account is not mentalistic. In fact, he advocates a naturalistic approach to intentionality: intentional states, processes, and events are for him

> part of our biological life history in the way that digestion, growth, and secretion of bile are part of our biological life history... Intentionality differs from other sorts of biological phenomena in that it has a logical structure. (1983, 160)

A consequence of this biological approach is to regard meaning – in the sense in which speakers mean something by their utterances – as a special development of more primitive forms of intentionality. In this sense, speakers' meaning is definable in terms of more basic intentional structures.[27] This is one reason why Searle understands the philosophy of language as a branch of the philosophy of mind. In his paper 'Intentionality and Its Place in Nature', for instance, he addresses the mind-body problem, claiming that there is no metaphysical gulf between the mind and the body – rather, just empirical ignorance about 'how it all works in the brain'. For him, the mental and the physical are not mutually exclusive categories; both 'are physical states of certain biochemical systems, or brains. But there is nothing reductive or eliminative about this view' (Searle 1984, 94).

---

25 Cf. Searle (1983, 1–3).

26 'On the contrary it seems to me obvious that infants and many animals that do not in any ordinary sense have a language or perform speech acts nonetheless have Intentional states. Only someone in the grip of a philosophical theory would deny that small babies can literally be said to want milk and that dogs want to be let out or believe that their master is at the door' (Searle 1983, 5).

27 This he also refers to as 'prelinguistic forms of intentionality' (ibid., 177).

The act of derivation

To understand more clearly how Searle uses the term 'derived intentionality', we must look at how he describes the act of derivation, or what he calls 'how to get from the physics to the semantics'. This is the question of how the mind imposes intentionality on entities that are not intrinsically intentional: 'Entities which are not intrinsically intentional can be made intentional by, so to speak, intentionally decreeing them to be so' (1983, 175). For Searle, 'linguistic meaning is a form of derived intentionality' (ibid.), and this marks an important distinction between the intentionality of mental states and that of speech acts: only where there is a 'distinction between Intentional content and the form of its externalization' (28) can we attribute meaning. Thus, we can ask for the meaning of a sentence but not for the meaning of a belief or other mental state. His general claim is that the mind imposes intentionality on the production of sounds, marks, and so on by transferring the conditions of satisfaction of the mental state onto the production of the physical phenomena. As already mentioned, Searle claims that there is a double level of intentionality in speech acts: the intentional state expressed, and the intention (in the ordinary, not technical, sense) with which the utterance is made. The second instance he calls the 'meaning intention'.

> Now it is the second intentional state, that is the intention with which the act is performed, that bestows the Intentionality on the physical phenomena. (1983, 28)

> […] the utterance act is performed with the intention that the utterance itself has conditions of satisfaction. (167)

One of his points of departure is Grice's claim that language is dependent on mind, in the sense that for speakers to mean something is to 'have a certain set of intentions directed at an actual or possible audience' (161). As we have already seen, Searle does not agree with the idea that this intention must be about producing effects in the audience. Rather, the necessary steps to get from the possession of intentional states to the performance of illocutionary acts, according to him, are (1) the deliberate expression of intentional states for the purpose of letting others know that one has them and (2) the performance of these acts for the achievement of the extralinguistic aims which illocutionary acts standardly serve.

He further lists several conditions of adequacy concerning the derivation of intentionality – for instance, the distinction between an intention to represent and an intention to communicate. While the latter is a matter of producing certain effects on one's audience, one can intend to represent something without caring at all about the effects on the audience. In Searle's account, representation is also prior to communication in the sense that there can be representation without communication, but not communication without representation. Furthermore, he delineates five categories of illocutionary acts: assertives, directives, commissives, declarations, and expressives. He claims that

> If the way that language represents the world is an extension and realization of the way the mind represents the world, then these five must derive from some fundamental features of the mind. (1983, 166)

This is another instance in which we can observe that he tries to explain how the mind works through the understanding of speech acts – and which gave me reason to speak of a perspective restraint.

Summary

We can summarize Searle's concept of derived intentionality in the following statements:

(d1) Derived intentionality applies to entities in the sense of an extension and realisation of the way the mind represents the world.

(d2) Entities with derived intentionality are representations in that their representational capacities are derived from the representational capacities of the corresponding mental states, and thus they necessarily express that mental state.

(r) A representation has a propositional content and a psychological mode, whereby the content determines the conditions of satisfaction and the mode its direction of fit.

(d3) The actual performance in which the speech act is made will involve the production (or use or presentation) of some physical entity, such as noises made through the mouth or marks on a paper.

(d4) It is the expression of a mental state with a certain set of intentions that converts the utterance act into an illocutionary act and thus imposes intentionality on the utterance.

3.3.2 Advantages of derived intentionality for Ref

I want to address two advantages that the concept of derived intentionality has for our understanding of the intentionality of artefacts and our emotional responses to fiction, namely, that it reinforces theaspect of presentation, and that it can help to address the question of reduced behavioural output.

3.3.2.1 The aspect of presentation

The main advantage of the concept of derived intentionality for understanding Ref is its substantiation of the claim that every fictional content is encountered in a form of presentation. Be it reading a book, watching a film, looking at a painting, attending a theatre piece, or listening to someone relating a story over a beer, we encounter a fictional content always in a form of presentation. What I earlier referred to as the *presentational aspect* of artefacts, is put by Searle in the following terms: every speech act has a propositional content and an illocutionary force. Furthermore, speech acts represent objects and states of affairs in the same sense of 'represent' as mental states represent objects and states of affairs, except that the intentionality of speech acts is derived, while that of mental states is intrinsic. Their structure, according to Searle is similar, though: while intentional states have a representative concern in a psychological mode, speech acts have a propositional content presented with illocutionary force. The psychological mode / illocutionary force is what makes a content be either an assertive, directive, expressive, or declarative. The mode of presentation is necessary for every mental state or speech act, which means thatwe have access to any intentional content only through a particular manner, and certain means of presentation.This is present in several aspects of his theory, but is best summarised in statement (d3):

(d3) The actual performance in which the speech act is made will involve the production (or use or presentation) of some physical entity, such as noises made through the mouth or marks on a paper.

Since we encounter fictional objects necessarily through artefacts, and if we understand artefacts to be entities bestowed with derived intentionality, this means that in the case of Ref (and not exclusively for them) our mental state is directed to something which is itself directed. Taking the presentational aspect of derived intentionality into account, which according to Searle involves 'the production or use or presentation of some physical entity', one can further argue that our mental state has access to the intentional content *through* the physical properties of this entity. This is the case for both fictional and factual accounts. Journal articles, documentaries, political speeches, medical imagery, and others – are all presented to us in a particular form, with a specific intention, and by certain means. To consider that these items are instances of derived intentionality acknowledges that there is necessarily an aspect of realisation, presentation, or use implied: that we have access to a certain content only through the form in which it is presented to us. As Searle points out, 'nothing is ever represented *tout court*, but only under some aspect or other' (1980a, 251).[28]

This consideration seems quite obvious. In our everyday encounter with fiction, we may often remark upon *how* a passage is written in a book we are reading, or *how* the camera moved at a crucial point in the film we have just watched; and we refer to such aspects not only out of professional interest. Rather, we point out how something is made because 'it' has moved us in a remarkable way. Furthermore, presentational aspects (sometimes also referred to as 'formal aspects' in aesthetics)are not only something we can appreciate in fictional accounts, but also in factual ones: we can be impressed by how a speech is delivered, or how a documentary is structured. Contrary to this everyday entanglement of our emotions with 'formalities', theoretical texts about Ref typically exclude the aspect of realisation of a work. Why is this?

The main reason, in my view, is the need of the modern debate around Ref to distance itself from the domain of aesthetics. The paradox focusses on fiction-directed intentionality, whereas debates in aesthetics are concerned with the appreciation of artworks, the definition of art, and so on. A paradox is constituted only (1) if our emotional responses are directed to fictional characters/situations, (2) if having emotional responses presupposes a belief in the existence of the object of emotion, and (3) if we believe that those fictional characters/situations do not exist. Excluding presentational aspects of a work is partly a way of ensuring that only cases that are paradoxical are addressed. In particular, the case that our emotional response is directed towards a presentational property of the work should be excluded. Choi (2003) puts forward this case in order to avoid the paradox. He argues that we have affective reactive responses elicited by the formal aspects of works of fiction. This specific case might avoid the problem the paradox poses; however, we still must account for

28 He develops this thought further in his *The Rediscovery of the Mind* (1992) and terms it 'aspect shape'.

cases of Ref which are experienced and formulated as directed to fictional characters and situations. Thus, the presentational aspect of works might have been neglected in order to exclude cases where the form of presentation is the object of the emotion. These are not cases that fall under the paradox.

For similar reasons, I differentiated emotional responses to fiction from emotional responses to art in the first chapter.[29] We must distinguish Ref from emotional responses to art (Rea) in order to focus on fiction-directed intentionality, and we must ensure that Ref are directed to a fictional content and not to its form of presentation. However, we must take into account that we always encounter fictions through artefacts, and that artefacts always convey content through their presentation, realisation, or use. In Radford's tragic-sister example, we can see that focussing on the nonexistent can make us blind to the importance of the presentational aspect of works. Being aware of this aspect concerns not only *how* something is presented to us, but (more importantly) *that* it is presented in the first place. With Ref, we are not directly confronted with a person or situation but rather with a representation. This concerns not primarily aesthetic questions but ontological and epistemological ones. We must not exclude the presentational aspect of works from our considerations about Ref, and understanding artefacts as items with derived intentionality helps to keep this in view.

The presentational aspect of works is not completely ignored in the debate around the paradox, as already mentioned with reference to Choi. I want to refer to a further example in order to show how the role of this aspect tends to be misunderstood in the debate. There is a tendency to assume that aspects of presentation detract from the coherence of the fictional world and thus inhibit emotional engagement with it. Kendall Walton calls the aspects that highlight the artificial character of a work 'ornamental features', and he gives examples such as van Gogh's conspicuous brushstrokes, or the 'flowery or alliterative' language of a novel (1990, 276). He states that such a feature 'may take on a life of its own, calling attention to itself at the expense of the thing described' (ibid.). Although there is good reason to distinguish emotions directed at fiction from emotions directed at presentational aspects of a work, we have seen that this does not justify the exclusion of the latter from our considerations. If Walton thinks that the form of presentation distracts our attention from the

29 I pointed out in 1.4 that questions of artistic value are not relevant for fiction-directed intentionality, and that the terms 'works of fiction' and 'works of art' are not coextensive.

fictional content presented to us, he is looking at the problem the wrong way around. How something is presented to us often determines whether a text or film captures our attention, and therefore potentially elicits our emotional engagement. Richard Moran objects to Walton along the same lines: 'we know that *Starry Night* would not really be more emotionally engaging if van Gogh had calmed down and left out all the overwrought brush work' (1994, 83). The matter is not one-sided, as the form of presentation can be used to different ends: it can help convey the fictional content and draw us into the story, but it can also be used to make us aware of the artificiality of the work. Its aim can sometimes even be to disrupt the 'illusion' in order to generate a critical distance to the presented content.[30]

A certain misunderstanding might also be due to a distinction made in the domain of aesthetics and art theory. There, representational, figurative or discursive aspects are often distinguished from so-called 'formal' ones, the latter referring, for example, to colour, composition, materiality, but also to techniques like montage in film, or other technical aspects of the making of a work.[31] I am referring to the presentational aspect of artefacts in the general sense that all artefacts are mediated: that they represent a content by means of the realisation, presentation, or use of a physical item or act. In these terms also the representational or figurative properties of a work are part of its form of presentation, because they are part of how a content is presented to us. Whereas the distinction in art theory is one internal to artefacts – differentiating different means of presentation – the distinction relevant to our concerns is one which differentiates artefacts from things in the world that do not have content. Furthermore, these domains might be using similar terms to different ends. While aesthetics is concerned with artistic value, formal means, and the aesthetic experience of Ref, the question relevant to the philosophy of mind is rather how it is possible *that* we have Ref in the first place, and what consequences this has for our concepts of fiction, emotion, and intentionality. The debate in the philosophy of mind is not restricted to artworks because, although most cases of fiction are artworks, it is not

30 Brecht's *Verfremdungseffekt* is perhaps the best known example, but only one of many.

31 These distinctions are not strictly delineated: they differ between different theories, and sometimes stem from a more colloquial use. Delineating these terms is not in within our scope here, but I aim to prevent confusion with my use of terms. Note that 'formal' or 'formalisation' have a quite different meaning in logic.

concerned with the question of what art is, but refers to any kind of artefact[32] that is representing fictional states of affairs.

To conclude: we should keep these questions separate, as our concern is to test our concept of intentionality in accounting for the phenomena of Ref. However, we must reconsider the importance of the presentational aspect for the discourse surrounding the paradox. The concept of derived intentionality emphasises that we are dealing with something 'made', that we bestow intentionality on a physical item, regardless of whether a fictional or factual state of affairs is represented. And in his concept of derived intentionality, Searle offers a useful distinction between representational content and illocutionary force, which he considers analogous to intentional content / psychological mode in mental states. This is not a question of meaning alone, but of individuation. Tim Crane claims that the content does not wholly individuate a mental state, but *how* we relate to it must also be specified. There is a difference between remembering a restaurant in St. Petersburg and imagining it. 'A person's intentional state is individuated by two things: the intentional mode and the intentional content' (Crane 2001a, 32).[33] Similarly, Searle explains the illocutionary force of speech acts by analogy with what he calls the intentional mode of mental states – we can claim, wish, or pretend that it is raining, while the propositional content remains the same. Thus, the presentational aspect is necessary for the individuation of the speech act: it disambiguates whether the utterance 'it rains' is a claim, a wish, or fictional discourse. Furthermore, the context, material, or style are part of how something is presented to us and can indicate whether we are dealing with a fictional or factual account, for instance. It is often not clearly stated in a text whether it is a fictional or factual account; rather, we infer this from aspects of its presentation.[34] The form of presentation cannot be excluded from the debate since it is our only access to the intentional content represented: the way in which content is presented to us is a crucial factor in its individuation. In the context of Ref, it is important to acknowledge that

32 This is why I distinguish the term 'artefacts' from works of art, and define it in 1.4 as 'physical items or performances with representational content'.

33 He further develops this thought in "The Intentional Structure of Consciousness" (2002).

34 We can recall the newspaper article about Jimmy, which Zipfel (2018) presented to his students: despite different points concerning the content, in the end they agreed that it must be a factual account because we encounter it in the *Washington Post*. However, that such aspects of presentation are indicative of fictionality/factuality does not mean that we should necessarily rely on such conventions.

we encounter fictions not directly but necessarily as mediated through a form of presentation. Thus, it is an advantage for explaining Ref that the concept of derived intentionality emphasizes the presentational aspect of artefacts.

3.3.2.2 Indirect behaviour

We are not dealing with a paradox properly speaking anymore since the problem of doxastic rationality has been solved; rather, according to Vendrell Ferran (2014), the debate revolves around issues of practical rationality now. One of the main concerns for this is that Ref typically do not lead to action, or only to reduced behaviours. We can be startled at the sudden appearance on the screen, but we do not run out of the cinema. Meskin and Weinberg (2003) determine 'reduced behavioural output' as one of the explananda that every theory concerning emotional responses to fiction must account for. This observation can be traced to the beginning of the debate, where one of the main points in Radford's tragic-sister example is to show that a decrease of emotional engagement occurs in the listener when he learns that the story is made up. Radford's intention is to argue that it makes a difference for our emotional engagement if we are dealing with a fictional or factual account of affairs. Let us recall the structure of the tragic-sister example and the two examples framing it:

(f-f) a seemingly factual account turns out to be false
(f-m) a seemingly factual account turns out to be made up (tragic sister)
(m-r) an instance of acting turns out to be based on real events

Radford constructs the three examples such that they display a similar structure: in the first two, a seemingly factual account turns out to be nonexistent (either false or made up), and in the third a seemingly fictional account turns out to be based on real events. His intention is to associate fiction with a decrease of emotional engagement, and fact with an increase. I have argued that the emotional response in Radford's examples is not due to the fictionality or factuality of events, but rather to the *turn* – the change in the logical or discursive status of an account. We are taken aback because the conditions of the account have changed and must reconsider how to relate to it. The decrease of emotion cannot be attributed to the fictionality of an account per se.

We have already discussed the topic of reduced behavioural output in Walton: he tried to establish it as a distinction criterion for Ref. I dismissed this, arguing that, firstly, other emotions also do not imply direct action and, secondly, that there are examples in which Ref lead to behavior.[35] Still, cases in which Ref display reduced behavioural output must be accounted for. The cases of Charles and Aaron could be explained by employing the unification view from the concept of affective intentionality. Perhaps the concept of derived intentionality can now help us address Henry's example. Henry is so taken by a theatrical performance that he wants to save the heroine; however, neither leaping onto the stage, nor pulling the plug of the sound equipment will do the trick.[36] His last idea is to change the plot:

> Henry makes it fictional that the heroine survives. He arranges things in such a way that this fictional truth is generated. But doing this is not saving the heroine [...] a painter or author can arrange for it to be fictional that an evil man dies, or that everyone lives happily ever after. But in doing so he does not kill the evil man or give everyone eternal bliss [...] one person can save another only if they live in the same world. (Walton 1990, 194–195)

Walton's point is that the interaction between worlds is not possible, that 'cross-world saving' is ruled out. If we change something 'in fact', it won't influence the fictional realm, and vice versa. Even if we change the plot, we don't really save the heroine; rather, it would become a different story, a new work of fiction.[37] Imagine coming across a version of *Romeo and Juliet* written by an ardent fan who rewrote the end with a better timing of the sleeping drug, and they live happily ever after. Few people would actually prefer such an outcome. Although we might be deeply moved by the fate of the heroine, we do not intervene in the performance because we know that we cannot really save her; and we (usually) do not rewrite the plot because we do not really want to change the story.[38] There is the common intuition that we have reduced behavioural output in such instances because we are dealing with artefacts. As was emphasised at the beginning of the chapter, we necessarily encounter fictional objects in a form of presentation: we are confronted with artefacts. This accounts for the lack of direct interaction. We know that we

35 The first was argued, for instance, in Moran (1994); the second, in Werner (2020).

36 In Walton's example the heroine is dying because she is bound to the rails of a train; in the last scene, the curtain closes and the sound of a train running over the stage is simulated.

37 Cf. Walton (1990, 193n2).

38 That we do not want to change the fate of the fictional characters despite empathising with them points to a different problem, namely the *paradox of tragedy*.

cannot directly interact with fictional characters and situations. Nonetheless, our behaviour might still be influenced by our emotional response to fiction. How our general behaviour is entangled with Ref is quite complex, and the examples discussed in the debate don't always do justice to this complexity.

The phrasing 'reduced behavioural output' focusses on the immediate interaction in a situation. To consider how our more general behaviour might be entangled with our emotional responses to fictional, let us consider the following examples:

(i) We encounter a homeless person on the street.
(ii) We read a pamphlet by Amnesty International about homeless people.
(iii) We are reading Molloy, a novel by Beckett, with a homeless person as protagonist.

Generally speaking, there is a considerably larger gap between (i) and (ii) than between (ii) and (iii). In the first case, we are directly confronted with something, and an action is not only triggered but often demanded by the situation. We might read about the same case of homelessness, perhaps looking at a picture of the same person as in (i) illustrating the pamphlet. However, we are not sharing the same physical situation; rather, we might be sitting at our kitchen table opening the post. It is obvious that a very different kind of (re)action is possible in the first two situations, although they might address the same person living on the street: while the first case allows for direct action, in the second case our behaviour can only be indirect. However, this does not mean that we would not act upon (ii). Albeit indirectly, we might have the same impulse to help, and we might act upon this impulse.

Scenarios (ii) and (iii) are both 'remote', but in the first we are confronted with a factual scenario; in the second, a fictional one. Nonetheless, both cases might influence our behaviour. Our empathy with Molloy might lead to a vaguer desire to somehow help,[39] whereas reading a pamphlet already addresses a very specific context and offers a concrete way to help through the projects of the organisation authoring the pamphlet. Although case (iii) might seem the most detached and abstract, it might bring about the biggest effect. It might make us relate to issues of homelessness in a more personal way, and we might

39 Obviously, this is a very reduced way of approaching Beckett's novel and does not in any way do justice to its artistic qualities (or even attempt to). Nor does choosing it as an example here suggest that the novel is 'about' homelessness.

look for possibilities to address the problem on a more structural level. We might research the structures that lead and sustain this problem – we might join an organisation or start a project ourselves. The impossibility of taking immediate action (such as donating money or food in that instance), and the lack of a prescribed option to help (such as supporting the organisation that produced the pamphlet), might in fact lead to a more structural and sustainable approach. Thus, despite the big differences between cases (i), (ii), and (iii), neither the level of emotional engagement nor the amount of action which might follow or continue our emotion can be categorically determined. There may be cases of (iii) that have a stronger emotional impact on us and thereby motivate us to take more significant action.

The homeless example shows that the entanglement of our behaviour with our emotional responses to fictional, factual, and real-life situations is quite complex. The examples typically put forward for 'reduced behavioural output' are very specific cases – such as being startled while watching a horror movie. They need to be accounted for but shouldn't make us forget the wider range of our emotional and behavioural entanglement with fictions. It has already been argued that reduced behavioural output cannot figure as a distinction criterion for Ref (see chapter 2.1 and 2.3.2). Considering different encounters with – and accounts of – homeless people, I showed how the phrasing 'reduced behavioural output' focusses on immediate (re)action. The concept of derived intentionality offers an advantage for explaining the behavioural output of our emotional responses to fiction because it keeps us aware of the fact that when we are dealing with fictions, we deal with representations. Consequently, our behaviour and actions towards fictional characters can only be indirect. This does not mean that such behaviour is weaker or stronger, reduced or 'full', more or less authentic. It is only not a direct reaction to the situation one is confronted with. The description 'reduced behavioural output' gives a reductive image of the phenomena addressed, as 'fits and startles' are very particular cases of Ref. Our behaviour is different when dealing with representations – fictional or factual – however, this should be better described as 'indirect'. The term 'indirect' only draws on the circumstance that any fictional content, and a lot of factual content, is mediated, and we are not inside the situation to react to it immediately. It does not have implications on the strength or the amount of action involved, and it does not allow us to draw conclusions from behavioural output to the authenticity of the emotional response involved. Thus, I propose to further speak of 'indirect behaviour' instead.

3.3.3 Disadvantages of derived intentionality

After pointing out some of the advantages of derived intentionality, I now want to look at two disadvantages the concepthas for a possible and appropriate understanding of the intentionality of artefacts, namely, the constitutive role of the author's intention in Searle's conception and the implications of the adjective 'derived'.

3.3.3.1 The author's intention[40]

One main characteristic of derived intentionality in Searle's understanding is that it 'is the intention with which the act is performed, that bestows the Intentionality on the physical phenomena' (1983, 28). In his conception, the mind imposes intentionality on an item/act by implementing the conditions of satisfaction of the mental state on the production of the physical phenomenon. This is happening through the intention with which the act is performed, which he calls the 'meaning intention'. Searle's point of departure is Paul Grice's claim that language is dependent on mind, in the sense that for speakers to mean something is for them to have a certain set of intentions directed at an actual or possible audience (Grice 1969). Searle disagrees with the idea that this intention must imply the production of effects in the audience, and he argues that 'meaning exists independently of the intention to communicate that meaning' (Searle 1986, 211). Rather, the conversion of an utterance into a speech act involves a 'meaning intention'. For Searle, the intentionality of artefacts is characterised by a 'double level of intentionality': the mental state which is expressed in the speech act and the intention with which the act is performed. Thus, the meaning intention is necessary for bestowing intentionality on a physical phenomenon. The act of derivation is summarised by him as follows:

(d4) It is the expression of a mental state with a certain set of intentions that converts the utterance act into an illocutionary act and thus imposes Intentionality on the utterance.

The idea that a sentence is the expression of a thought and that, to express or represent something, we need to have the intention to do so seems quite intuitive. Although he focussed on speech acts, he attributed derived intentionality

40 An earlier version of this section was published as Nistor, Iulia (2024). 'Unintentional Intentionality', Humanities Bulletin 7 (1): 9–16.

to a wider range of phenomena, including writing and pictures (Searle 1980a). Furthermore, the emphasis on the meaning intention helps explain different types of illocutionary acts, such as assertions, directives, commissives, declarations, and expressives. The plain representation of a mental state would be, for instance, an expressive which does not yet imply any intention to communicate, but only an intention to express. So the 'set of intentions' can explain the propositional attitude, for instance, if a content is presented as a judgement, desire, or order.[41]

However, if we stop in our analysis here, we are considering only one perspective: that of the speaker. Searle is looking at the phenomenon of intentionality through the perspective of speech acts. What I earlier called *the perspective restraint* now becomes relevant. His perspective is that of a speaker, and his conception of derived intentionality is shaped by this restraint. I am going to argue that one disadvantage of the concept of derived intentionality is that it cannot account for all phenomena of intentionality of artefacts. The problem becomes apparent in the meaning intention – or 'author's intention'.[42] For Searle, the author's intention is a necessary condition for derived intentionality: it is what bestows intentionality on the physical phenomenon. He is explaining the act of derivation from the perspective of the author, more specifically the speaker, and it seems quite clear, if we put ourselves in the position of a lecturer giving a seminar or a speech in front of his students. However, since we are discussing emotional responses to fiction, we must also be concerned with the perspective of the readerof a novel or the audience in a cinema. Looking at it from this position, things become more complicated: how can we understand a sentence, or image? And what is the role of intentionality herein? Being confronted with a speaker, we can simply ask him about his intentions, if they were not clear from his lecture. However, in most other cases when we are confronted with artefacts – when reading a book, watching a film, or looking at a painting – the author is usually not present, might refuse to explain his intentions, or is dead. But we might still find meaningful the words and images we are confronted with. They seem to convey, or give us access to, content. In order to account for the intentionality of artefacts we must also consider thereceiver's point of view, which does not necessarily have access to the author's intention. Furthermore, we must account for images, not only text. I

41 Cf. Searle (1983, 166ff).

42 I am using the term 'author's intention' as synonymous with 'meaning intention' in order to emphasise that Searle focusses on the perspective of the speaker.

am going to argue that intention is not necessary for the intentionality of an artefact, and show that it is a disadvantage of Searle's notion of derived intentionality that it binds intentionality so tightly to the author's intention. There are cases of intentionality in the absence of intention, which I will call 'unintentional intentionality'.

To discuss unintentional intentionality, we begin with Steven Knapp and Walter Benn Michaels' wave poem example (1982): walking along the beach, we come across the stanza of a poem written in the sand. We can read and understand the writing, and we may assume that someone has written it there. Eventually, a wave washes over it and astonishingly leaves another stanza behind, which seems to continue the poem. Knapp and Michaels concluded that while in the first case, we assume that the sentences have meaning, or semantic content, because we suppose the existence of an author – and his intention – in the second case, we do not attribute meaning to the signs in the sand. As there is no author, but the marks appear accidentally, for them, the second stanza is not a linguistic phenomenon. They proceed to consider sentences produced by a computer, arguing that meaning in this case depends on whether we can attribute intentions to computers and not on the possibility of 'intentionless language' (Knapp and Michaels 1982, 728). Searle's position is referenced in their account, introduced with the sentence 'The claim that all meanings are intentional is not, of course, an unfamiliar one in contemporary philosophy of language' (727). In response, Searle offers a further example: suppose a cat wanders over the keys of a computer keyboard and out comes the perfect English sentence 'The chair is made of wood'. He argues that nothing prevents us from taking this instance as an English sentence and attributing meaning to it since

> Languages are human institutions and thus are intentionalistic through and through. But it is a mistake to confuse the intentionality of the system with the requirement that every element of the system be the product of an actual human intention. (Searle 1994, 680)

Searle claimed that Knapp and Michaels misunderstood his point. In his view, a sentence is the standing possibility of an intentional speech act. Conversely, Knapp and Michaels believe that without the instantiation of the author's intention, nothing is a sentence. I further suggest that Knapp and Michaels seemingly confused intentionality with intention. In the passages mentioned earlier, they debate the possibility of an 'intentionless language' and the claim that 'all meanings are intentional'. In neither case is it clear if they mean 'intentional' in the technical sense. Furthermore, there is no instance in the text in which intentionality is explicitly addressed in a technical sense, such as through paraphrase. It seems that, for them, the 'intentionality of language' means the 'intendedness of a sentence or speech act'. This is not implausible, given that Knapp and Michaels background is in literary theory, in which the author's intention is widely discussed.[43] Even in philosophy, it is uncommon to attribute the term intentionality to writing and utterances. Rather, one would speak of reference or representation, whereas intentionality is typically attributed to states of mind. Searle applied the concept of intentionality to physical phenomena, such as writings, utterances, and pictures, referring to it as derived intentionality. For him, 'linguistic meaning is a form of derived intentionality' (1983, 175). For now, I will use 'meaning' – or something 'being meaningful' – based on Searle's sense of something 'having derived intentionality', but I will differentiate this later.

Instances of unintentional intentionality, such as the wave poem or the cat-on-the-keyboard example, are unproblematic for Searle. According to him, we understand a correct sentence in a language familiar to us,

> weaned from all production or origin, putative or otherwise. […] because a meaningful sentence is just a standing possibility of the corresponding (intentional) speech act. To understand it, it is necessary to know that anyone who said it and meant it would be performing that speech act determined by the rules of the language that gives the sentence its meaning in the first place. (1977, 202)

The rules of language and the standing possibility of an author's intention ensure intentionality in such cases. This suggests that not every instance of language – that is, not every written sentence or speech act – must be made with the intention of an author, but it is the standing intention presupposed by readers that makes the ascription of meaning possible. The reader does not necessarily need access to the particular intention with which the act is performed,

---

43 However, even with respect to the author's intention, we have to differentiate between the discourses. Whether the author's intention should be a norm for the meaning of a work, or is necessary for any kind of meaning to exist, is a different question. I could insist on the importance of the author's intention (and the context of production, etc.) for the purpose of interpretation by an art historian or an art critic, parallel to the argument that intention is not a necessary condition for intentionality in the philosophy of mind.

nor does there need to be a particular intention involved in the first place. This allows for examples such as the wave poem or the cat on the keyboard. In these cases, there is no intention on the part of the author (or the cat),[44] or indeed no author in the first place (waves washing over the sand), but in both such accidental agglomerations of signs can appear meaningful if they form a (more or less) correct sentence in a familiar language. It seems that the conditions for intentionality differ according to the author's perspective and that of the reader or listener. From the author's perspective, the necessary steps to get from the possession of an intentional state to the performance of an illocutionary act, according to Searle, are (a1) the expression of that intentional state according to the rules of the language used, and (a2) the performance of the illocutionary act with the intention of an author. From the perspective of the reader, the understanding of a sentence is constituted by (r1) the reader being familiar with the rules of the language used and (r2) the reader presupposing the standing possibility of the speech act being performed with the intention of an author.

Clearly, languages are human 'institutions' – systems that function according to a set of rules – which are essentially based on intentionality. This implies that in order to impose intentionality on a physical item/event, one must (a1) (per)form an item/event that fulfils the rules of language, and the intentionality of such an item/event can be understood if (r1) the audience is firm in the rules of (that) language. However, it is unclear why the intentionality of physical phenomena must necessarily be linked to an author's intention, even if we understand this as the standing possibility of such an intention. According to Searle, (a2) is necessary because there is a difference between saying something and meaning it and saying something without meaning it.

> [People] make noises, or marks on paper; they draw pictures, or wave their arms about etc. Now my problem is: what must be added to these noises, marks, etc., in order that they should be statements, orders, etc.? What, so to speak, must be added to the physics to get the semantics? (1986, 209)

However, the thing 'added' does not seem to be a matter of how something gathers semantic content but rather of how something becomes a speech act. There is a difference between uttering a sentence and meaning it: for example, singing a song in a foreign language. We can sing a sentence meaning it and

44 The cat might have several intentions for walking over the keyboard, but (probably) not the intention to write the sentence 'The chair is made of wood'.

without meaning – or understanding – it. Searle is interested in this difference. He asks how noises become a statement, an order, or the like. Therefore, the 'meaning intention' seems to refer to what we mean to do by making these noises. However, although a sung sentence may not be meant, it still has meaning. If, for example, our friend Katrin is singing along to a song on the radio – *a cadeeeeiiira e feita de madeeiiiira* – she might not mean to say that 'the chair is out of wood'. She may not even know the meaning of the words she is singing. And we might not understand them either because we do not speak Portuguese or because of her pronunciation. Nevertheless, this is a correct Portuguese sentence, and a Portuguese speaker would be able to understand it. Therefore, the difference in this case is what Katrin means to do, that is, singing along to a song or making a statement, not what the sentence means.

Catherine Talmage argued that Searle failed to distinguish clearly between 'meaning' in the sense of the semantic meaning of a sentence and 'meaning' in the sense of a speaker meaning to do something by uttering that sentence.[45] One can say something without meaning to, which nevertheless has meaning. In Katrin's case, we could still doubt whether she meant only to sing along to the song or (also) to make a statement. Let us suppose that Katrin pronounces the sentence 'The chair is made of wood' while fast asleep. She articulates a correct English sentence, and we understand it. We would not hold her responsible for what she said because we do not consider a person in that situation an agent. However, what she utters forms a correct sentence nonetheless, and nothing prevents us from understanding it. She does not mean what she says – that is, her utterance is not a speech act – but the uttered sounds can be considered an instance of intentionality because they form a correct sentence and thus seem to be about something. Consequently, (a2) might be true for illocutionary acts but not for all instances of intentionality.

Let us look at (r2). Is it necessary from the perspective of the reader, listener, or observer to presuppose the (standing) intention of an author to understand a physical phenomenon as an instance of intentionality? In most cases, there is a reason to presuppose an author, and in these cases, readers and audiences assume the intention of such an author and often wonder about the particular intention she had. When we read a sentence from a book, an advertisement on a billboard, or graffiti on a wall, we usually assume an author

45 Cf. Talmage (1996).

and her intention.[46] Even when we read a poem in the sand – as described by Knapp and Michaels – we initially assume an author's intention. However, after a wave washes over it and leaves behind another poem, we are certain of witnessing an accidental agglomeration of signs in the sand. Still, these signs make sense to us, and nothing prevents us from understanding them. It seems necessary that the accidental agglomeration of signs forms a correct sentence in a language we are familiar with. However, is it necessary for their understanding to presuppose (r2) the standing possibility of an author's intention? What could this 'standing intention' actually be? There are many cases where the author is not present, and we do not witness her performance. For example, Searle (1977) mentions the sentence of a dead author: 'On the twentieth of September 1793, I set out on a journey from London to Oxford'. In this case, the standing intention could be related to the production of the sentence and the context in which it is read. When we read a sentence in a book, we assume that there was an author who intended to make such a statement because it is a printed sentence in a book. Someone must have decided to print it; before that, someone must have submitted it for publication; and before that, someone must have written or typed it. When we read the sentence scratched in the seat in front of us while sitting on a train, we assume the author and her intention because someone must have meant to scratch it there. However, in the case of the wave poem, we just saw the wave wash over the first stanza and leave behind a second stanza, so we would not assume that it had been written there by someone. We are astonished by this coincidence and bemused that the random signs make sense to us. It is precisely the absence of an author (and her intention) that elicits such bemusement. What could a standing intention be in this case? Perhaps that the sentence could be used by someone in a speech act – that is, the possibility of appropriating the signs for an author's intention. However, in this case, (r2) is not necessary to understand the signs; rather, it is because someone understands the signs that they might appropriate them. In a more general sense, the standing intention might be inherent to the conventions of language; it could be the presupposition that, at some point, someone attributed a certain meaning to a certain sign and that this was done intentionally. This presupposition may be implicit in our use of language. In this sense, (r2) would be true for strongly conventional signs – like most words we use in language.

46 Actually, this assumption is becoming less and less common, given the increasing presence of AI-generated images and texts.

(r2) may be true for texts and other manifestations of language. However, if we now consider images, we find this changes the setup significantly. Let us assume that our wave does not leave behind a poem but a clearly delineated portrait of a particular person – say, Madonna. We do not follow the lines of the drawing as we would follow the words that form a sentence, but the image is more instantly recognisable to us. At least at the level of 'face value'. Also an image can be highly coded when used in a manner that implies complex contexts and conventions.[47] We even sometimes speak of having to 'read' an image or of a 'language of images'. In this case, I do not discuss such coded uses of images but rather the basic recognition of an object or person. We have said that (r2) might be true for strongly conventional signs because the intention of an author can be understood as implicit in the arbitrary conventions of language. With regard to images, this setup is changing; Images are not an arbitrary system agreed upon by convention, and, in their case, it would be absurd to assume that at some point, someone (intentionally) attributed a certain meaning to a certain sign. Such an assumption might be unrealistic – but not impossible – in the case of language.[48] Images partly function through the recognition of their likeness to the things they depict, whereas the words we use to signify things usually do not bear a likeness to the things they signify.[49] Symbols, signs, and pictograms may be exceptions and must be discussed separately. The crucial difference for my argument is the arbitrariness of linguistic meaning. This helped maintain (r2) earlier: that the reader presupposes the standing possibility of the speech act being performed with the intention of some author. Moreover, images have conventions of use and may be highly coded. However, in the case of the wave portrait, we recognise Madonnabecause the signs in the sand resemble her. We do not presuppose that at some point someone intentionally attributed the meaning 'Madonna' to that particular sign. Thus, images form an exception – as (r2) applies only to arbitrary signs.

The phenomenon of seeing images in accidental stimuli is also referred to as pareidolia – such as recognizing shapes in clouds, seeing a man (alterna-

47 The study of iconography, for instance, is dedicated to this.

48 The notion that the conventions of language originate in an intentional act of denomination by some author(s) is merely a hypothetical assumption that a reader bring to a text, and does not pertain to the actual formation of the conventions of language.

49 Likeness cannot define images (for a critique of likeness theories of images, cf. Oliver Scholz, 2004); however, it does form an important difference to the arbitrariness of linguistic signs.

tively, a rabbit) in the moon, or the face of the Virgin Mary on a piece of toast. Scientific research considers pareidolia a sensory illusion because, in attending to an actual stimulus, a familiar pattern is perceived. The same could be said of representations, only we usually assume that someone made them with some purpose. The tendency to recognise faces in accidental stimuli is particularly pronounced. Recent neuroscientific research has shown that our sensory system works similarly if confronted with (what is called) an illusionary face or a real one. For example, we track the gaze of illusionary faces in the same manner as we track the eyes of a person before us (Takahashi and Watanabe, 2015). Several studies have investigated pareidolia using functional magnetic resonance imaging (fMRI), and the results indicate, among other things, that the interpretation of accidental stimuli depends on processes similar to those elicited by real objects. This is related to the finding that such recognition occurs at a relatively early stage of perception and is not a late cognitive reinterpretation of sense data (Voss et al., 2012). Some researchers have suggested that our sensitivity to face recognition and its location in early sensory processes may also explain why the mere perception of a face can produce the same reactions we have when we look at a real person (Kaufmann et al., 2018). The fact that image recognition is not a late cognitive reinterpretation of sense data but can occur at an early stage of perception supports our theoretical observations that (r2) – the presupposition of a standing intention – is not necessary for intentionality.

One could argue instead that the viewer's intention is necessary for intentionality – that the viewer wants to see a face – as in games of seeing-in, where one tries to find shapes in the clouds passing by. We sometimes play such games and what we want to see (or read) certainly influences the meaning we attribute to that image or sign. Furthermore, what a viewer is familiar with surely determines what he is able to recognize. If someone had never seen Madonna (or a picture of her), they would not be able to recognise her face in the sand. They might only vaguely recognise a face, or perhaps it would remind them of their sister or of someone else they know. What different viewers recognise is contingent upon their experiences. Recognition presupposes correspondence with something seen previously. However, this does not imply that the viewer's intention is necessary for intentionality. This is because recognising something is not (usually) a wilful act but happens spontaneously. Except for in games of seeing-in, we typically do not intend to recognise something. Rather, it happens 'by itself': a certain similarity strikes us, which may come as a surprise. To recognise a face on a piece of toast comes as a surprise:

firstly, because it is an unlikely place for an image to appear, and, secondly, because we know that it is an accidental agglomeration of signs. Still, we recognise an image. We might dismiss the thought as absurd – we might laugh about it and not take it seriously. However, we can only experience this reaction because we recognise something in the first place. This means that, in this case, the intention of the viewer can be excluded.

Discussing cases of unintentional intentionality and considering not only text but also images, we can see that (a2) may be true for illocutionary acts but not for all instances of intentionality. As, for Searle, the phenomenal range of derived intentionality includes – next to utterances and writings – also pictures,[50] it follows that intention is not a necessary condition for intentionality.

3.3.3.2 The adjective 'derived'

The second difficulty we encounter in the concept of derived intentionality is related to the very core of the term, namely, to the adjective 'derived'. This adjective indicates, according to Searle, the relation between the intentionality of artefacts and the intentionality of the mind.

'Derivation' is, according to the *Cambridge Dictionary*,'the origin of something, such as a word, from which another form has developed, or the new form itself'. It can describe the etymology of a word as well as the change of its meaning or grammatical category. The application of the term is fairly broad: in linguistics it means 'the formation of a word from another word or base'; it can mean 'source or origin' of foods, or 'descent' in the sense of ancestry; and in logic it is considered 'a necessary consequence of previously accepted statements'. In which sense is Searle's use to be understood?

Searle is concerned with the question of how the mind imposes intentionality on entities that are not intrinsically intentional: he asks 'how to get from the physics to the semantics'. His general claim is that the mind imposes intentionality on the production of sounds, marks, and so on by imposing the conditions of satisfaction of the mental state on the production of the physical phenomena. The necessary steps to get from the possession of intentional states to the performance of illocutionary acts, according to him, are (1) the expression of intentional states, in which the psychological state expressed in the performance figures as the sincerity condition for the speech act and (2)

50 Cf. 'intentionality applies […] to linguistic entities such as speech acts and sentences, not to mention maps, diagrams, laundry lists, pictures, and a host of other things' (Searle 1983, 26).

the performance of these acts for the achievement of the extra-linguistic aims, which makes the illocutionary act an assertive, directive, commissive, declaration, or expressive. According to him, an account of meaning must explain how the conditions of satisfaction of the speech act are determined by the conditions of satisfaction of the sincerity condition such that they are identical. He summarises that the intentionality of the mind 'creates the possibility of meaning, [and] it limits its forms' (Searle 1983, 166), particularly referring to the functions speech acts derive from intentionality.

We can distinguish two ways in which Searle might be using the term 'derived'. On the one hand, he tries to explain how the conditions of satisfaction of the sincerity condition can be imposed on a physical item, that is, how the content of a mental state can be securely derived in a speech act. For a content to be securely derived from a mental state the conditions of satisfaction of both the mental state and the physical item must be identical. We can call this *content derivation*. On the other hand, he understands the capacity to represent as derived. He claims that

> the main function which language derives from intentionality is, obviously, its capacity to represent. Entities which are not intrinsically intentional can be made intentional by, so to speak, intentionally decreeing them to be so. (1983, 175)

This means to understand derivation as the imposition of a function – the function to represent – on a physical item. We can call this *function derivation*. Searle does not make this distinction, but he understands these two aspects as interdependent: the imposition of representational function depends on the derivation of content, and vice versa. This is justified for the majority of cases; however, I think that it is important to clarify the specific mechanisms involved. For this reason, I argue that we must differentiate these two aspects of derivation, and I will contend that the capacity to represent is independent of whether the sincerity condition is fulfilled in the performance of the speech act. To support this claim, I will examine cases of 'diverging intentionality'.

As an example of diverging intentionality, we can recall Jastrow's duck-rabbit,[51] a simple drawing which can be seen either as a duck or a rabbit. Let us presuppose the author intended to draw a duck. What the viewer sees, however, is

51 Joseph Jastrow introduces this image in his book *Fact and Fable in Psychology* (1900). The example was famously picked up by Wittgenstein and has often been referenced since.

a rabbit. The author imposed the conditions of satisfaction (the shape of a duck) of the sincerity condition (the idea of the duck) on a physical item (lines on a piece of paper). However, the viewer sees a rabbit. In this case, the intention of the author did not unequivocally determine the intentional properties of the artefact. For the author it is a drawing of a duck; for the viewer it is a drawing of a rabbit. The content of the mental state ('duck') was securely imposed on a physical item by the author, but a different content ('rabbit') is derived from the properties of the artefact by the viewer. Does this depend on the viewer not having access to the intentions of the author? I don't think so. Let us imagine a case in which the author is telling us about his intention and is perhaps even pointing out how we must look at the drawing in order to see the duck. We are able to see the duck now, but we can still skip back to our first impression and see the rabbit. So we can attribute a diverging intentionality despite our knowledge of the author's intention. The possibility of allowing for both readings – that is, to be justified in deriving both the content 'duck' and the content 'rabbit' from the drawing – is to claim that the author intended to draw a duck-rabbit. In this case we could say that the viewer is justified in deriving both contents because both were part of the artist's intention. This, however, would still not guarantee 'secure' content derivation. The sincerity condition is only met if the conditions of satisfaction of the mental state and the physical item are identical, that is, if the viewer sees a duck-rabbit in this case. It is possible, however, that some viewers see only a duck, and others see only a rabbit.

One might argue that these difficulties stem from the fact that we brought in the viewer. The author's derivation, that is, the making of an artefact according to the mental state he intends to express, has to be distinguished from the derivation of a viewer. Here, anaudience derives the thought content from the sentences read, or the images seen. Audiences do not have 'direct' access to the thought contents of the author, but are confronted with an artefact, that is, with a form of presentation. Lamarque, for instance, says that 'mental representations or thought-contents [are] identifiable through descriptions derived in suitable ways from the propositional contents of […] sentences' (1981, 302). However, the question of meaning is very complex. How we derive meanings from artefacts and whether these must correspond with the intentions of the author constitute the subject of entire domains. Searle focuses on the speaker and how an utterance can be made into an illocutionary act. In this case, 'derivation' primarily refers to the act of imposing a thought content on a physical item.

Can we think of a case of diverging intentionality that does not involve an audience but the author alone? An instance of such a case would be when someone tries to express a mental state in a speech act but fails to do so. Here, a mental state is expressed in the performance of the speech act with a certain intention. However, although the author has tried to meet the sincerity condition of the thought content in his performance of the speech act, he has not quite achieved it to his satisfaction. One of Searle's points of departure is Grice's claim that language is dependent on mind, in the sense that for speakers to mean something is to 'have a certain set of intentions directed at an actual or possible audience' (Searle 1983, 161). Because Searle does not agree with the idea that this intention has to be about producing effects in the audience, he distinguishes between representation and communication: one can intend to represent something without caring about the effects it has on one's audience. It must be possible to lie, and it must be possible to make a statement while failing to make a true statement, or to be indifferent to whether the audience believes or understands it. Thus, on Searle's account there can be representation without communication, but not communication without representation.

We can add to this that it must be possible to make a statement while failing to make a sincere statement. That is, it must be possible to make a statement and not fulfil the sincerity condition of the mental state we are trying to express, perhaps because it is very subtle, or simply because we cannot find the right words in the moment. This case is not necessarily related to how an audience understands a statement. We might try to express a certain thought content in writing a sentence and realise, on reading it, that it does not quite express the thought in question.Let's say we try to describe how Peking duck tastes. We try hard to find the right words, because it is our favourite dish. When we read the sentences we wrote, we realise, however, that the words fail to convey the mental content we tried to express. What we wrote forms a correct sentence, which we can understand and attribute meaning to, but it does not exactly convey what we wanted it to express. The description does not meet the sincerity condition of the mental state: we made a statement but failed to make a sincere statement. This case of diverging intentionality can help us distinguish the two types of derivation mentioned earlier. In this case the function derivation is successful, but the content derivation is not: we have created a meaningful sentence but failed to meet the sincerity condition of the mental state we intended to express. This is nothing extraordinary, but happens all the time to scholars, poets, or indeed anyone who tries to express a particular thought. We just cross out the attempt and try to find better words. However, the example shows that the capacity to represent is independent of whether the sincerity condition is met in the performance of the speech act. Thus, it allows us to differentiate function derivation from content derivation.[52] The speaker successfully bestowed intentionality on a physical item while failing to impose the exact conditions of satisfaction of the content he wanted to express.

If we recall Searle's quote that the intentionality of the mind 'creates the possibility of meaning, [and] it *limits* its forms' (1983, 166, my emphasis), we could take this as the weaker claim of derived intentionality. The Peking duck example qualifies as such a weak form of derived intentionality, because there is a meaning conveyed in the sentence written, and the description might even come fairly close to it; however, it is not the exact meaning intended. The strong claim would be that the intentionality of the mind creates the possibility of meaning, and *determines* its forms, as claimed in (d2):

(d2) Entities with derived intentionality are representations in that their representational capacities are derived from the representational capacities of the corresponding mental states, and thus they necessarily express that mental state.

My emphasis on the 'that' points to the fact that function derivation here implies content derivation: not any, or a similar content is expressed, but *that* exact one.With the Peking duck example, we saw that content derivation does not necessarily follow from function derivation. The representational capacities of the artefact are derived from the representational capacities of the mental state, but this does not imply that the artefact necessarily expresses that exact mental state. At least not in the sense of 'secure' derivation: that we can derive *that* particular content from a sentence or speech act. We might be able to give a close description of how Peking duck tastes, such that it can be distinguished from a rabbit stew, for example. However, it might not convey the exact thought content we wanted to express and thus does not precisely individuate that content. Searle persisted in arguing the strong claim because he tried to explain how speech acts work. He uses the concept of intentionality to ground his speech act theory. However, with examples of diverging intentionality we saw that the strong claim of content derivation cannot be upheld for all cases of derived intentionality. Moreover, whether an audience derives thought contents 'in suitable ways' from artefacts is rather a question of meaning.

52 Similar as in the distinction between representation and communication, there can be function derivation without content derivation, but not vice versa.

It is important to distinguish between intentionality and meaning at this point. As I mentioned earlier, Searle claims that linguistic meaning is a form of derived intentionality. However, this should not imply that intentionality and meaning are identical.[53] Rather, they stand in an asymmetrical relationship: if there is meaning, there must be intentionality, but not vice versa. We can identify intentional properties in/on physical phenomena, such as an agglomeration of marks that form a correct sentence or bear a likeness to the features of a person. However, although we might say that these marks have an intentional property, they only have meaning if an author intends to express a meaning through them, or if a viewer reads them as meaningful. In this sense intentionality is not identical with meaning; rather, the intentional properties of physical phenomena should be described as potentially meaningful.

While the questions of meaning and intentionality overlap, they are not identical. The issue of intentionality is primarily a matter of how physical items or events can be said to have the capacity to represent, and how they acquire it – that is, how a semantic or intentional property can be attributed to something in the physical realm. How can this capacity, typically associated with mental states, be conferred onto physical phenomena? For Searle to say that a speech act is a representation is simply to say that it has a propositional content and an illocutionary force. The derivation of the *function* to represent can be understood through this structure: a physical item becomes a representation when it is assigned both propositional content and a presentational force. However, while every representation must have *some* content, it does not need to have a particular content to fulfil a representational function. Hence, I distinguished between *function derivation* – that something comes to serve as a representation – and *content derivation* – what that representation is specifically about.

Furthermore, it is essential to distinguish between intentionality and intensionality, as these two notions are frequently conflated. Intensionality is neither a criterion for, nor a sufficient condition of, intentionality. Intentionality refers to the directedness or aboutness of mental states – our mind's capacity to be directed towards objects, properties, or states of affairs. Intensionality is a linguistic or logical notion. A sentence is said to be intensional when

53 Searle equates linguistic meaning with derived intentionality but distinguishes it from intrinsic intentionality. For him, meaning is part of the disanalogy of mental states and speech acts. Only speech acts (and other artefacts) can have meaning; mental states can't. Cf. Searle (1983, 175)

it fails standard tests for extensionality, such as the substitutability of identicals and existential generalisation. Accordingly, intentionality is a property of minds, while intensionality is a property of sentences or linguistic contexts. In intensional contexts, logical properties are sensitive to the way things are described, rather than to their referents or extensions. For instance, in belief reports like 'Lois Lane believes that Superman can fly', substitution of coreferential terms (e.g., 'Superman' and 'Clark Kent') may alter the truth-value of the sentence. This reflects intensionality, but it does not follow that belief itself is an intensional phenomenon in the same sense. Some sentences about mental states create intensional contexts, but not all intensional contexts involve mental states, and not all reports of mental states are intensional. As Searle puts it, the mistaken belief that intentionality is inherently intensional arises from the 'confusion of features of reports with features of the things reported' (1983, 24). This confusion can also be traced back to Roderick Chisholm, who helped introduce the concept of intentionality into the analytic tradition by defining it in terms of intensional criteria. This, as Tim Crane argues, encouraged a linguistic interpretation of intentionality, where intentionality was seen as a semantic property of language, with mental states deriving their representational power from linguistic structures. To distinguish these two notions – while acknowledging their connection – Crane proposes that 'there are two kinds of directedness, or two ways of reporting directedness: the intensional and the extensional' (2001a, 18). This maps onto the classic *de dicto* / *de re* distinction. In this light, one of the most problematic aspects of applying the concept of intentionality – which properly describes a capacity of the mind – to speech acts, sentences, or other artefacts, is the persistent confusion between intentionality and intensionality.

If we intend to attribute derived intentionality to artefacts, we have to, on the one hand, carefully differentiate it from notions of meaning and intensionality and, on the other hand, understand the intentional properties of artefacts as derived from the intentionality of mind in the weak sense of function derivation: that the intentionality of the mind creates the possibility of meaning and *limits* its forms.

3.4 Intentional properties

Is it helpful to attribute intentional properties to artefacts, and in what sense can such properties be understood? To address the former, we must try clarifying the latter.Both mental states and artefacts share the property of being directed to, of being 'about' something. However, how can intentional properties be individuated in the case of physical items and events?

First, we must distinguish *function derivation* from *content derivation*. If we understand artefacts as physical items or events with representational content, then we may attribute derived intentional properties to all artefacts in the weak sense of function derivation, but only to some in the strong sense of content derivation. This distinction accounts for cases of diverging intentionality, as presented earlier. In many cases, an author expresses a mental state through a text, performance, or image, thereby imposing intentionality onto a physical phenomenon. These are examples of illocutionary acts. Here, the intentionality of the artefact is derived from the intentionality of a mental state through the author's intention. This is the case Searle seeks to account for: the conditions of satisfaction ensure that the artefact is 'about' the same thing that the expressed mental state is 'about'. However, there are also instances of mistakes and unsuccessful attempts. These cases do not lack intention; rather, the presence of intention confirms the failure. We can identify something as a mistake only if there was an intention specifying a desired outcome in thefirst place. Furthermore, it must be possible to make a statement and fail to make a sincere statement – that is, fail to fulfil the sincerity condition of the mental state we try to express. In the Peking duck example, we try to describe how Peking duck tastes, but upon reading our own description realise that we have failed: we created a meaningful sentence but failed to meet the sincerity condition of the mental state we intended to express. Such cases of diverging intentionality show that the capacity to represent is independent of whether the sincerity condition is met in the expression of intentional content. It therefore allows us to distinguish function derivation from content derivation. While meaning is concerned with content derivation, this need not apply to intentionality. Function derivation refers not to *what* something represents but *that* it represents. It does not require that the content of the artefact be 'securely' derived from a mental state, or that audiences be able to derive contents in 'suitable ways' from the artefact. Rather, an artefact derives the capacity to represent from an author's intentionality, in the sense that the intentionality of the mind creates the 'possibility of meaning and limits its forms'. This is how we can distinguish function derivation from content derivation. If intentional properties are to be attributed to artefacts, only function derivation necessarily applies.

Secondly, we must distinguish between *derived* and *underived* intentional properties. In the process of converting thought content into a sentence, image, or other artefact, we express a thought when we form a sentence about it or produce an image capturing it. Artefacts can represent our thoughts in such a way that they are directed at the same content as the corresponding mental state. In this sense, the intentionality of artefacts may be derived from the intentionality of mind (either that of the author or of a reader/ listener). Yet we must also account for instances of intentionality in the physical realm that are not derived – namely, instances of unintentional intentionality, as discussed in the example of pareidolia. Here,we are confronted with accidental agglomerations of signs or marks. Despite the lack of an author's intention (regular or standing), we are able to understand a text, or to recognise an image in these configurations. Such instances cannot be considered artefacts, but still they seem to represent something. We might attribute intentional properties to such instances in the sense that they exhibit the capacity to be 'about' or 'of' something. A piece of toast that resembles an image of Madonna does so because it bears a certain coincidental similarity to her – not every arbitrary configuration of marks produces this effect. Furthermore, such accidental instances can be *used as* artefacts when someone recognizes their representational potential. Even if the toast's resemblance occurred by chance, it can subsequently be used intentionally *as* a representation of Madonna. However, if we attribute intentional properties in such instances, these must be qualified as 'underived'.

Searle developed the concept of derived intentionality to ground his speech act theory. This is evident in the adjective 'derived', which describes the relationship between the intentionality of the mind and that of physical phenomena: the latter is derived from the former. According to Searle, the intention with which an act is performed imposes intentionality on physical phenomena. Yet, as shown above, intention is not a necessary condition for all cases of intentionality. Therefore, the dependency of intentional properties in physical items or events on the intentionality of the mind must be called into question. Such dependency may be necessary for meaning (i.e., content derivation) but not for intentionality. When we encounter physical instances of intentionality, we cannot infer that they were made by an author and derived from a corresponding mental state. We can only ascertain that both mental states and physical phenomena can instantiate intentionality – but not that one must necessarily derive from the other. Importantly, attributing underived intentional properties to physical items does not entail ascribing intrinsic intentionality to them.

This brings me to the third distinction: the one between intentionality and intentional properties. I will develop this distinction in the next chapter. For now, I only stipulate it: there is a difference between intentionality as an *activity of the mind* and intentionality as a *property of mental states*. This distinction allows us to attribute both derived and underived intentional properties to physical items and events, while maintaining that intentionality as an activity is exclusively a capacity of the mind. Searle's distinction between intrinsic and derived intentionality might suggest that 'underived' is synonymous with 'intrinsic', but this would be misleading. By 'underived intentional properties', I mean only that these properties are not derived from a prior mental state; this does not imply that they are instances of what Searle calls intrinsic intentionality.

Encountering instances of intentionality in different realms – in mental states, and in physical items or events – does not necessitate that they are derived from one another. This would entail assuming a source or origin of intentionality. In the case of artefacts, we have the tendency to search for an origin to securely determine their content. While this may be important for questions of meaning, it does not bear the same relevance for intentionality. As the cases of diverging and unintentional intentionality demonstrate, an origin is neither always available, nor necessary for the individuation of intentional properties. Although content derivation may be necessary for questions of meaning – that is, for determining the appropriate or intended meaning of an artefact – it does not hold the same importance for intentionality. The tendency to extend this requirement to intentionality is what I refer to as the 'fallacy of origin'.

If we distinguish intentional properties from (intrinsic) intentionality, and if we separate function derivation from content derivation, then we can attribute intentional properties – derived by an author/audience or accidentally occurring (underived) – also to physical items and events. Based on these distinctions, we can define an intentional property as a property that 'indicates contentfulness':

(ip) An intentional property is a property that indicates contentfulness.

This definition emphasises not the particular content but the *having* of content – the capacity to be about something, rather than what something is about. I use content instead of object, following Crane, to allow for the possibility of (1) nonexistent objects and (2) the distinction between what one is thinking about and what one is thinking. This further helps distinguish intentional properties from reference: while the object of a sentence may be understood as its referent and serve as the truth-bearer of that sentence,[54] the content of a sentence is what is expressed by that sentence. This is not to say that attributing an intentional property is denoting the content of the sentence, but rather *that* the sentence has a content – that it indicates content. It is worth noting that Brentano already distinguished intentionality from reference: for him, all mental states are intentional, but not all refer to something. Reference only occurs when the object of the mental state exists in reality.[55] Thus, attributing intentional properties does not necessarily entail an existential commitment regarding the object(s) of the intentional relation.

There is something about the toast that looks like something else, that points beyond itself: it indicates content. In this sense, the toast can be said to have intentional properties. That these properties emerged by coincidence does not affect the matter. Furthermore, this property can be 'read' in different ways: someone might vaguely recognize a face, while someone else might recognize their sister, and another person might say it is a spitting image of Madonna. The intentionality of artefacts is only potentially meaningful; it still requires a viewer or reader to attribute meaning to them. To speak of intentional properties in such instances, however, allows us to address the fact that something *seems* to be 'about' something. This can help explain the potential function of such items or events: that we are able to attribute meaning to them or use them in a meaningful way, e.g. by appropriating them. However, this remains a potentiality. We can disregard the resemblance to a face that seems to appear on a piece of toast and see it as a mere agglomeration of marks. The same holds for more conventional signs: we can, for instance, see a written word only as a shape, disregarding the meaning to which it is conventionally tied.[56]

---

54 However, the object of a sentence might be also understood as its grammatical object, which would not involve an existential commitment.

55 This was pointed out by Hamid Taieb (2017). Brentano does not use the word 'reference' but speaks of a relation of 'similarity', which in Taieb's analysis amounts to reference. He puts forward the following quote: 'The fact that every thinking being has an object, a content, does not mean that it is a relative faced with a correlative. It is directed toward something as an object. When the latter exists, a kind of relation exists, which we could call a similarity in a peculiarly modified sense, between the being that is taken as an object and the thinking being' (Brentano 1952–1966, 324).

56 That we are able to see a word or text only as a shape is, for instance, crucial for concrete poetry. In this case, the shape of a text codetermines the meaning, and we can oscillate between what is conveyed by the words and what the shape of the text resembles.

Is it helpful to attribute intentional properties also to physical items and events? One reason we may be discouraged from doing so is what I refer to as the *fallacy of origin* – the tendency to individuate a content by appealing to its source or origin. However, as we have seen, an origin is neither always available nor necessary for individuating an instance of intentionality. Most importantly, attributing intentional properties to physical phenomena should not lead us to conflate intentionality with intensionality. I have distinguished intentionality from both meaning and reference, and further differentiated between content derivation and function derivation, between derived and underived properties, and between intentional properties and (intrinsic) intentionality. With these distinctions in place, we can understand intentional properties as properties that *indicate contentfulness*. In this sense, it would at least not be incorrect to attribute intentional properties to certain physical items and events.

3.5 Conclusion

How can we now address question (B)" whetherthere is a difference in intentionality if it is mediated or immediately directed at a state of affairs? We can consider three options:

ms ––––– o

ms ––––– a ––––– o

ms ––––– a (o)

I suspect that our encounter with artefacts is often understood in the first sense: that when we read a book or watch a film, our mental states are directly related to, for instance, Frankenstein's monster. The intentionality of artefacts was not problematized, and the form of presentation was rarely taken into account in the context of the paradox. Only in the first case would the answer to (B)" be 'no'. In both other options, there is a difference. Searle's understanding of derived intentionality is an advantage in so far as it acknowledges that the intentional object is mediated through a physical item or event. Thus, we can attribute the second option to Searle: the intentional object is mediated through a form of presentation. As we saw in the discussion of Radford's tragic sister example, any major change or break in this setting – such as a shift in the logical status of the account – might have more impact on us than the account's fictionality or factuality itself. The third possibility, which I am opting for, essentially follows the second model but emphasises that the intentional content is 'fixed' in a form of presentation. In this case, the content is not merely mediated through a form of presentation; rather, it is fixed within that form. This means that we do not have direct access to the intentional content but only to its form of presentation: we are confronted with the intentional properties of an artefact, and these do not guarantee secure content derivation.

Due to what I call the *fallacy of origin* – the tendency to individuate a content by appealing to its origin – we assume that we can securely derive the content of an artefact, for instance, from the intention of its author. The main reason why I am opting for the third possibility is because of the distinction between function and content derivation. We may attribute derived intentional properties to all artefacts in the weak sense of function derivation, but only to some in the strong sense of content derivation. This means that when we encounter intentional properties in physical items, they indicate that they are vehicles of content; however, this does not guarantee that we gain access to the exact content intended. The function of representing requires a form of presentation, and the possibility of meaning is created by limiting its forms. However, since artefacts are fixed in a form, their content becomes ambiguous, as we do not have direct access to the intentional object – only to its form of presentation. That is, it cannot be assumed that the exact content intended by the author can be derived, or even that the author succeeded in conferring the intended content on the item. So, whereas in the second case it seems that we just have to get the meaning right, in the third case the gulf between mental state and represented content is insisted upon.

If we decide to attribute intentional properties also to physical items or events, we must pay attention to what these properties are meant to denote. They would have to denote something other than either meaning or reference. To this end, I proposed defining intentional properties as 'properties that indicate content'. To illustrate this, let us consider the description *a(o)*, with *a* standing for the physical item or event and *o* for the intentional object or content. The content can be related in two ways: intensionally and extensionally. That is, *o* may, depending on the context, be understood as the reference or the meaning of the artefact. According to the proposed definition, the notion of 'intentional property' denotes neither meaning nor reference. Rather, it refers to the parenthesis *(...)*: that which indicates that an item has content. In this sense, the notion captures the condition of something having the function, or being able to serve the function, of representing; that something, for instance, constitutes a correct sentence that we can understand – regardless of what it means or whether someone actually wrote it. Whether introducing such a notion proves useful remains to be seen.

4. The 'general' understanding of intentionality

> And then sometimes there arose within me, confusedly, a kind of consciousness, which I express by saying, I said, etc., or, Don't do it Molloy [or] by means of other figures quite as deceitful, as for example, It seemed to me that, etc., or, I had the impression that, etc., for it seemed to me nothing at all, and I had no impression of any kind, but simply somewhere something had changed, so that I too had to change, or the world too had to change, in order for nothing to be changed. And it was these little adjustments, as between Galileo's vessels, that I can only express by saying, I feared that, or I hoped that, or, Is that your mother's name? said the sergeant, for example, and that I might doubtless have expressed otherwise and better, if I had gone to the trouble.
> *Samuel Beckett, Molloy*

Having examined intentionality through the two perspectives offered by the paradox of fiction and considered the concepts of affective intentionality and derived intentionality respectively, I now turn to what I have previously referred to as the 'general' understanding of intentionality. The discussions in the preceding chapters revealed that, regardless of the specific conceptions – whether affective or derived – there are different ways in which *intentionality itself* is understood and used. This is perhaps the most significant, and in some respects, concerning outcome of my analysis so far.

I set out to use the paradox of fiction as a tool to analyse the concept of intentionality. What has emerged is a landscape marked by divergent, and potentially incompatible, conceptions of the term. These different 'general' understandings of intentionality must be taken seriously. Some of them are explicitly identified in the literature – such as 'propositional intentionality', which is increasingly regarded as overly narrow.Others have required clarification, for instance, the distinction between'objectual intentionality' and 'directedness',pointed out in chapter2. Moreover, intentionality, as originally conceived by Brentano, is a complex notion and was never entirely unambiguous in its initial formulation.

The question now arises whether these various general understandings merely highlight different aspects of the same phenomenon, or whether they point to a case of conceptual equivocation. It is conceivable that the divergent interpretations correspond to different traditions of thought, notably to how the term evolved in the analytic and continental realm. While a historical analysis of this development would undoubtedly be both informative and necessary, such an investigation lies beyond the scope of this work. Instead, I pursue a different strategy: I examine the various paraphrases used to describe intentionality. As 'intentionality' is a technical term, authors often use paraphrases when introducing or referring to it. Such paraphrases can be indicative of potential conceptual differences. In this chapter I analyse some of these paraphrases in order to determine whether the implicit understandings of intentionality that we encountered in the last chapters are fundamentally distinct. In particular, I focus on the descriptions of intentionality as *content and directedness*, and as *property or capacity*. These descriptions will provide the structure for this chapter. I begin with a brief history of the term, followed by the discussion of intentionality as content and directedness, with particular attention to Brentano's original formulation and its subsequent reception. I then turn to the understanding of intentionality as property or capacity,drawing on a further divergence found in the proponents of affective intentionality. Finally, I show that these two notions of intentionality are already present in Brentano's writings without having been explicitly differentiated by him.

4.1 Paraphrases and origin of the term 'intentionality'

Since 'intentionality' is a technical term, we are usually given a less technical paraphrase or description when it is first mentioned or introduced in a text.

In lexica and anthologies, intentionality is variously described as the 'mind's capacity to direct itself on things' (*Routledge Encyclopedia*), as 'directedness, aboutness, or representational properties of mental states' (Lycan, 2008), as the 'directedness of a psychical act or a state of affairs' (*Historisches Wörterbuch der Philosophie*, 1976, my translation),[1] or as the 'power of minds and mental states to be about, to represent, or to stand for, things, properties and states of affairs' (*Stanford Encyclopedia*). Furthermore, neologisms such as 'of-ness' or 'about-ness' are often used in contemporary texts as synonyms for intentionality. Among these different descriptions, it seems that on the one hand, 'having a content' and 'being directed towards' are used synonymously. On the other hand, intentionality is sometimes called a 'property of mental states' and sometimes a 'capacity of the mind'. These paraphrases correspond to different aspects of the understanding of intentionality in the authors discussed in the previous chapters. Are these paraphrases compatible, merely highlighting different aspects of the same phenomenon, or do they indicate that fundamentally different understandings of intentionality are at play?

I will first briefly outline the history of the term, its origins, and its introduction into philosophy. Etymologically, it is derived from the Latin *intendere* ('to stretch out', 'to focus on, turn attention towards', 'to direct, aim', 'to intend'), and it is here still connected with the meaning 'intention/intent'. In medieval Latin, *intendere* is also used with the meanings 'to understand', 'to think, believe', or 'to go to, travel', and it was in medieval philosophy that the term *intentio* was first specified and used as a technical term. It appears as a translation of the Arabic words *ma'qual* and *ma'na* in the works of Al-farabi and Avicenna. Al-farabi uses *ma'qual* in his commentary on Aristotle's *De Interpretatione* to refer to the concept of *noema*, differentiating 'something in the soul denoting both what is outside it, [...] "prima intentio", and what is inside the soul, a "secunda intentio"' (Gyekye 1971, 35). Both terms were further used by Avicenna and Al-Farabi, respectively, to determine the subject matter of scholastic logic:

> The subject-matter of logic – as you know – consists in the second intelligible intentions (or concepts, al-ma'qulat) which are dependent upon the first intelligible intentions (al-ma'qulat).[2]

1 The German original reads 'Gerichtetheit eines psychischen Aktes auf einen Sachverhalt'.

2 Avicenna (*al-Ilahiyyat*, 10); citation from Gyekye (1971, 36).

> As for the subject-matters (al-maudu'at) of logic [...] they are the concepts (or the intelligibles: al-ma'qulat), inasmuch as they are signified by words, and words inasmuch as they signify concepts.[3]

The two words *ma'qual* and *ma'na* were translated into medieval Latin as *intentio*, which was then used as a technical term. Gyekye points out that there has sometimes been confusion between the ordinary and the technical sense of the term in the translations from Arabic to Latin and from Latin to modern languages – for example, in translating *prima intentio* as 'primarily' or *principaliter*, rather than in the scholastic sense of the term.[4] *Intentio* in a technical sense was described, for example, by Ockham as 'something in the soul which is a sign, naturally signifying something for which it can stand for or which can be part of a mental proposition'.[5] However, the scholastic tradition was mostly interested in the distinction between thing – *intentio* – word, and the relation between *primae* and *secundae intentiones*, wherein *intentio* was not properly defined but was generally understood as 'that which is immediately before the mind' (Knudsen 1982, 479).

It was Brentano who reintroduced the concept to modern philosophy. While in medieval philosophy the term *intentio* was used, Brentano spoke of *intentional in-existence*, similarly drawing on Aristotelian thought. He describes the concept in a threefold way: the intentional in-existence of a mental phenomenon is

> what we might call, though not wholly unambiguously, reference to a content [Beziehung auf einen Inhalt], direction towards an object,[...] or immanent objectivity [immanente Gegenständlichkeit]. (Brentano 1994, 68 [1924, 124])[6]

There are two things to remark upon here. Firstly, the ambiguity of the term – which he points out himself – and which I will elaborate on in the following chapter. Secondly, that neither in the medieval tradition nor in Brentano is

3 Al-Farabi (*Ihsa al-ulum*, 1949, 59); citation from Gyekye (1971, 36).

4 Cf. Gyekye (1971, 24–25). He also refers to the technical sense as the 'scholastic' or 'Avicenna-sense' of *intentio*.

5 As quoted in Gyekye (1971, 34).

6 I am giving here and in the following the page of the English translation and the respective page in the German edition edited by Kraus, eventually adding the German terms in brackets if relevant.

'intentionality' used in its substantive form. In the 20th century and up to the present, the concept has mostly been referred to by the term 'intentionality', but Brentano himself used it only as an adjective – as in *intentional in-existence*, or in saying that mental states 'intentionally contain objects'. The term became prominent in the phenomenological tradition but was – somewhat independently – also increasingly used in the analytic realm, where there have been various attempts to reduce or explicate it. Furthermore, in the analytic tradition intentionality was often understood in terms of propositional attitudes, that is, as a relation between a subject and a propositional content. I will elaborate further on the reception of the term in the analytic tradition later.

Intentionality also came to be widely used and applied across various domains. As we saw in chapter 3, Searle applied the concept to speech acts, distinguishing between a primary or intrinsic intentionality of the mind and a secondary or derived intentionality of language. According to this view, cultural artefacts like films, books, images, and sentences do not mean something in themselves, but through the intentionality of the mind. I examined some of the advantages and disadvantages of this approach. In chapter 2, I demonstrated the particular importance of intentionality for emotion theory. Here, intentionality serves different functions: notably, it provides the emotional experience with its 'about', thus making emotions amenable for rational assessment. Further, it was used to differentiate intentional emotions from nonintentional feelings. Emotions are different from affective reflexes or patterns of behaviour because an involvement with their content is part of their motivation. Döring and Reisenzein, in this vein, point out that 'thanks to their representational content, the emotions are capable of making other mental states and actions rational, rather than merely causing them' (2009, 201). Historically, intentionality also helped integrate emotions into wider philosophical debates, that is, what is referred to as the *cognitivist turn* in emotion theory.

This short introduction is intended to provide only a sense of orientation regarding where the term 'intentionality' originally comes from, as well as its vastness and complexity of use across various domains. There is an extensive literature ranging from its ancient origins, through a thorough reading of Brentano and the different stages of his theory, the attempts to explicate and naturalise it in the analytic tradition, up to the recent differentiation of different kinds of intentionality, such as phenomenal, non-propositional, or affective intentionality. It is, of course, impossible to review and encompass all this research here. However, given the different – often implicit – understandings encountered in the previous chapters, it seems crucial to gain a more fundamental grasp of the concept. But how can this be approached? To navigate this vast domain, I will structure my analysis around the paraphrases most commonly used for the term. I propose identifying (some) of the paraphrases of intentionality as *content* and *directedness*, as *property of mental states* or as *capacity of the mind*. These paraphrases are reflected in the various implicit understandings of the authors discussed in this study. The differences we have encountered must be taken seriously and made explicit. I will examine whether the use of these different paraphrases is compatible, or whether fundamentally different understandings of intentionality are at play – that is, whether they refer to aspects of the same phenomenon or denote different phenomena altogether.

4.2 Intentionality as content and directedness

I am first going to refer to the paraphrasing of intentionality as content and directedness. I will show that these descriptions can already be found in Brentano by analysing a prominent passage where he describes his notion of intentional inexistence. In order to clarify the paraphrases of intentionality as content and directedness, I will first analyse Brentano's use of the term, look at its reception, and point out its mutual explanation.

4.2.1 Franz Brentano's notion of intentional in-existence

In the central passage which describes the notion of *intentional in-existence*, Brentano points out a certain ambiguity herein, by stating that he is 'not wholly unambiguously' describing this notion. I want to look at this prominent passage for different reasons: first, to get a proper understanding of his notion *intentional in-existence*, but also to discern what exactly he is referring to as ambiguous. The passage also contains two footnotes where he explains his particular understanding of existence and refers to examples from the scholastic and ancient tradition.

> Every mental phenomenon is characterized by what the Scholastics of the Middle Ages called the intentional (or mental)* in-existence of an object, and what we might call, though not wholly unambiguously, reference to a content [Beziehung auf einen Inhalt], direction toward an object (which is not to be understood here as meaning a thing), or immanent objectivity [im-

> manente Gegenständlichkeit]. Every mental phenomenon includes something as object within itself, although they do not all do so in the same way. In presentation something is presented, in judgement something is affirmed or denied, in love loved, in hate hated, in desire desired and so on.** This intentional in-existence is characteristic exclusively of mental phenomena. No physical phenomenon exhibits anything like it. We can, therefore, define mental phenomena by saying that they are those phenomena which contain an object intentionally within themselves.
>
> * They also use the expression "to exist as an object (objectively) in something", [gegenständlich/objektiv in etwas sein] which, if we wanted to use it at the present time, would be considered, on the contrary, as a designation of a real existence outside the mind. At least this is what is suggested by the expression "to exist immanently as an object" [immanent gegenständlich sein], which is occasionally used in a similar sense, and in which the term 'immanent' should obviously rule out the misunderstanding which is to be feared.
>
> ** Aristotle himself spoke of this mental in-existence [Einwohnung]. In his book on the soul he says that the sensed object, as such, is in the sensing subject; that the sense contains the sensed object without its matter; that the object which is thought is in the thinking intellect. In Philo, likewise, we find the doctrine of mental existence and in-existence. However, since he confuses them with existence in the proper sense of the word, he reaches his contradictory doctrine of the logos and Ideas. The same is true of the Neoplatonists. St. Augustine in his doctrine of the verbum mentis and of its inner origin touches upon the same fact. St. Anselm does the same in his famous ontological argument; many people have observed that his consideration of mental existence as a true existence is at the basis of his paralogism (cp. Überweg, *Geschichte der Philosophie*, II). St. Thomas Aquinas teaches that the object which is thought is intentionally in the thinking subject, the object which is loved in the person who loves, the object which is desired in the person desiring, and he uses this for theological purposes. When the Scriptures speak of an indwelling of the Holy Ghost, St. Thomas explains it as an intentional indwelling through love. In addition, he attempted to find, through the intentional in-existence in the acts of thinking and loving, a certain analogy for the mystery of the Trinity and the procession ad intra of the Word and the Spirit. (Brentano 1995, 68–69 [1924, 124–125])

The threefold description of the notion *intentional in-existence* is followed by a further explanation of how Brentano thinks that mental states contain something within themselves. This is helpful for clarifying how we can understand intentional in-existence nowadays. What remains unclear is how the different descriptions that he used to introduce this term relate to each other. To determine what he means by ambiguity, we must examine the different descriptions and see how they relate. The descriptions are 'reference to a content' (Beziehung auf einen Inhalt), 'direction towards an object' and 'immanent objectivity' (immanente Gegenständlichkeit). The grammatical structure of the sentence could be read as a row of different descriptions with the structure $d_1 \wedge d_2 \wedge d_3$ or it could be read as the first two being used synonymously and the third being distinguished from them, which would form the structure $(d_1 \wedge d_2) \vee d_3$. By ambiguity, Brentano might mean that all these descriptions indicate different aspects of intentionality, or that the first two descriptions together refer to one aspect and the third to another.

The first two descriptions are closer in meaning because they both suggest a direction. Reference implies a direction in that it names or signifies something it is not – it is pointing beyond itself or, as etymologically implied, 'carrying back' to something. Translating *Beziehung* as 'reference' might be misleading because Brentano did not mean to imply the strong sense of *direct reference*, for which reference fails if the referent doesn't exist. By 'Beziehung auf einen Inhalt', he only means that something is presented in thought, that a mental state is 'about' something, without making an existential commitment.[7] The following description, 'direction towards an object', can be seen as paraphrasing this. It might be meant to clarify that this relation/reference is understood as directedness and does not imply the existence of its referent. At a different point, Brentano also uses the description 'reference to something as an object [Beziehung auf etwas als Objekt]' (1995, 74 [1925, 137]). The phrasing 'as an object' here specifies that he is talking about grammatical objects, not physical ones. In this sense, the descriptions d_1 and d_2 seem to be almost interchangeable. One reason why he is using both might be in order to distinguish between what the scholastics called *prima intentio* and *secunda intentio*, denoting

7 Cf. the important difference between intentionality and reference, which has been pointed out by Taieb (2017), and I mentioned in 3.4. Taieb proposes to translate 'Beziehung auf einen Inhalt' in the mentioned passage as 'relation to a content' (2017, 122). This, however, does not avoid the existential commitment suggested, since the logical notion of relation similarly requires both relata to exist. Whether translated as 'reference' or as 'relation', we have to keep in mind that mental presentation for Brentano does not involve an existential commitment.

something outside or inside the mind, respectively – that is, if we take 'content' to mean 'mental content', and 'object' to mean 'physical object'.[8] The third description has different implications: the adjective 'immanent' indicates that something is contained rather than directed. Brentano attached a footnote to further clarify this meaning. The term *immanent* is supposed to prevent the misunderstanding of a physical existence outside of the mind. Furthermore, *objectivity* is not meant to be understood as opposed to subjectivity but rather as a translation of the German *Gegenständlichkeit*. It does not denote the aspect of physical existence, but that something 'exist[s] *as* an object (objectively) in something' (1995, 67, my emphasis). This could be translated as 'object-liness' or 'object-ivity' and describes a property of mental states. In a second footnote, Brentano further grounds the notion of immanent objectivity by explaining that 'Aristotle himself spoke of this mental in-existence. In his book on the soul, he says that the sensed object, as such, is in the sensing subject; that the sense contains the sensed object without its matter' (ibid.). Compared to the first two descriptions, d_3 displays a different aspect of Brentano's conception of intentionality. Whereas d_1 and d_2 are related to the idea of directedness, d_3 is more related to that of (mental) content. The neologism 'in-existence' reflects this aspect of Brentano's understanding, and it even became part of his main notion for the concept of intentionality: *intentional in-existence*. In German, it is also referred to as *Einwohnung*, which can be translated as 'in-dwelling'. It does not designate non- or quasi-existence but that a mental state has something as its object, namely, what is presented, judged, loved/hated, desired, and so forth. The notion of existence remained problematic, and he referred to it again in an annotation in the appendix from 1911, explaining that it is a loose use of the term, which we allow out of convenience:

> In many cases, the things to which we refer do not exist. But we are accustomed to saying that they then have being as objects. This is a loose [uneigentlicher] use of the verb '"o be", which we permit with impunity for the sake of convenience, just as we allow ourselves to speak of the sun "rising" and "setting". All it means is that a mentally active subject is referring to those things. […] Because reference to things is a distinctive characteristic of someone who is mentally active, we have been led to talk about objects having being or subsisting in the mentally active subject. Similarly, the fact that

8 He explicitly states in brackets that 'object' here is not to be understood as 'meaning a thing', that is, a physical object, which makes such a reading unlikely.

> the subject refers to the same thing in different ways has led people to talk about something which is in some way more than the object, for it includes the latter within itself, and is likewise within the subject. It has been called the "content" of the mental reference. Especially in connection with the mental act of making judgments, there has been talk of a content of judgement as well as an object. If I judge, "a centaur does not exist", it is said that the object is a centaur, but that the content of the judgement is that a centaur does not exist. (Brentano 1995, 227–228 [1925, 158–159])

Brentano often uses the terms 'object' and 'content' interchangeably but differentiates them at a certain point to imply that the content can include further qualification of the object. The two terms are not coextensive because we can refer to the same term in different ways. This can be understood similarly to propositional attitudes in language. Brentano himself uses the example 'a centaur does not exist' to mark the difference between object (the centaur) and content (that centaurs do not exist), pointing out that a mental object exists as the 'content of the mental reference' (ibid.). As pointed out earlier, 'object' here is used to mean a grammatical object and does not involve an existential commitment. Brentano is generally not as concerned as we are nowadays with how something in the mind can relate to something in the world; rather, his concern is to delineate the realm of the mental. The description of *intentional in-existence* as 'immanent objectivity' (*Gegenständlichkeit*) is best understood as the claim that mental phenomena contain intentional objects within themselves in the sense of 'mental content'.

To conclude my analysis of this central passage for Brentano's understanding of intentionality: first, Brentano is not using the term intentionality but the adjective 'intentional', and he is mostly speaking of 'intentional in-existence'. Further, I consider that the notion of *intentional in-existence* can be best understood in the sense of mental content. Finally, I conclude that, with the phrase 'not wholly unambiguously', Brentano is referring to the understanding of intentionality as content and directedness. The first two formulations ('reference to a content' and 'direction towards an object') specify the kind of relation mental phenomena have to their objects and can be redescribed with the term 'directedness'. Whereas the third, 'immanent objectivity', is concerned with how a mental state can have a (grammatical) object and can be redescribed as 'content'. Consequently, the structure of the sentence should be understood as $(d_1 \wedge d_2) \vee d_3$.

Furthermore, it appears that he needed both content and directedness in order to describe his concept of intentionality. This indicates what I will call the *twofold nature of intentionality*. I will come back to this aspect after looking at the reception of Brentano.

4.2.2 The reception of Brentano

Brentano's purpose in introducing the concept of intentionality differs from the questions concerning intentionality nowadays. Tim Crane claims that Brentano's notion of 'intentionality' differs from ours. He points out, as I mentioned earlier, that Brentano introduced the term to delineate the mental as the realm of psychology, and that for him intentionality is the mark of the mental (Crane, 1998). He further contrasts our understanding of intentionality as the 'of-ness' or 'about-ness' of mental states with Brentano's notion of intentional in-existence (Crane, 2017). To argue that there was a shift in the use of the term, Crane closely examines the reception of Brentano in analytic philosophy. I want to present his considerations briefly.

Crane agrees with Barry Smith, who claims that 'one will find no coherent interpretation of Brentano's concept of intentionality as long as one remains in the framework of our usual commonsensical notions of the mind and its objects' (Smith 1994, 40). Smith argues that we have to consider a certain background of beliefs, what he calls 'commonsense realism', that we have nowadays: this view assumes that there are things that exist independently of our mind, that relations hold between these things, and that science is studying these things. Crane explains how Brentano's understanding is a different one. One cannot say that Brentano is a phenomenologist, because he believes in a reality beyond phenomena; however, for his point of view phenomenology concerning the sciences could as well be true because the sciences can be only concerned with phenomena. This is why Peter Simons also calls him a methodological phenomenologist.[9]

Next to this different worldview, Crane further points out the reception of Brentano in the English-speaking realm. He claims that there has been a wrong interpretation of Brentano's theories, and foremost of his concept of intentionality. This is, on the one hand, due to the fact that the first translation into English was made only in 1973 and is based on Kraus's edition from 1924, which means that few knew Brentano's writings in the original. Also, according to

9 Cf. Simons's introduction to Brentano's *Psychology* (1995).

Crane (2017, 41), Kraus's footnotes and annotations gave further cause for misunderstandings.[10] On the other hand, Brentano's concept of intentionality was mainly introduced to analytic philosophy by Roderick Chisholm.[11] Chisholm tried to apply Brentano's concept to the differentiation of sentences which describe mental phenomena and those which describe physical phenomena, and thus to prove the irreducibility of the former. Crane claims that on this basis other (analytic) philosophers came to understand intentionality as the having of a propositional attitude, and he gives Quine and Davidson as examples.[12] He points out that Brentano's theory aims at determining a distinction criterion between physical and mental phenomena, and is not concerned with the structures of language (nor would Brentano have believed in propositional attitudes in a 20th-century sense). According to Crane, these considerations regarding the reception of the term 'intentionality' explain the shift in its meaning.

For Crane, the focus in Brentano's concept of intentionality lies in its understanding as 'immanent objectivity' (*Gegenständlichkeit*). By the 'objectivity' of mental phenomena, Brentano denotes their 'having something as a content', wherein the object is immanent in the mental act. This means that the notion of *in-existence* must be taken literally in Brentano, because the object is immanent and does not lie beyond the thought. Thus, Crane gives priority to the notion of 'immanent objectivity' (*Gegenständlichkeit*), and the descriptions 'reference to a content' (*Beziehung auf einen Inhalt*) and 'direction towards an object' are used in order to explain how a mental state can 'contain an object intentionally within [itself]'. It is important to be aware of the different worldviews during Brentano's time and today, and that Brentano used the term for purposes different from those used in the current discourse around intentionality. Furthermore, Crane's research on the reception of the term in the English-speaking realm is of interest for my considerations, and I am going to come back to it in my argumentation later. Crane's emphasis on the understanding of intentionality as immanent objectivity (*Gegenständlichkeit*) might be, to a certain extent, a reaction to the shift in meaning the term underwent, an attempt to recover

10 Cf. Crane's introduction to Brentano's *Psychology* (2015).

11 Prominently in his text 'Sentences about Believing' (1955).

12 Quine, for example, believes that Brentano was reviving the scholastic term intentionality to describe verbs of propositional attitude(*Word and Object*, 1960). This assumption is brought further by Davidson in *Mental Events* (1970) in claiming that the characteristic of mental verbs is that these express propositional attitudes (cf. Crane 2017, 49f).

an aspect which has been lost. In my consideration of the relevant passages of Brentano's *Psychology* I could not observe a priority between the three descriptions: 'reference to a content', 'direction towards an object', and 'immanent objectivity'. Although the structure of the sentence, as we saw, should be understood as $(d_1 \wedge d_2) \vee d_3$ the first two descriptions do not amount to an explanation of the third. Rather, the first two together describe the sense of directedness in intentionality, while the third concerns its aspect of content. However, content and directedness seem to depend on one another, and Brentano seems to need both aspects in order to explain his concept of intentionality. This brings me to my point about the *twofold nature of intentionality*.

4.2.3 Mutual explanation of content and directedness

For the understanding of intentionality as content and directedness, the terms 'object' and 'relation' are of particular interest, and the way in which Brentano uses them is interesting: he relativises and explains each with its respective other. I have already provided some explanations of the term 'object'. When we think about something, we would say that this mental state has an object or that it has something as its object.[13] According to Brentano, this is a loose way of speaking and does not imply the existence of such an object, either physically or mentally. He strongly disagrees with the belief in the existence or subsistence of mental objects as displayed in the theories of (his pupils) Husserl and Meinong, for instance. In the essay 'On Ens Rationis', dated 1917 and added to the appendix of the 1924 edition, he writes:

> Even in the modern period a number of such non-things have been spawned by people who are always going around accusing others of psychologism. They imagined they were enriching science by their many weighty discoveries. Gegenstandstheorie or, as they called it, Phenomenology, was proclaimed to be a special science entirely independent of psychology. They revived the error of Plato and the ultra-realists like William of Champeaux with certain modifications, by ascribing a being to universals as universals. This discovery was supposed to open the way to the true metaphysics, and,

13 Brentano sometimes explicitly formulates the intentional relation as 'Beziehung auf etwas *als* Objekt' (1924, 137, my emphasis), and Kraus has pointed out that 'object' is used by Brentano as shorthand for 'to have something as an object'. Cf. Kraus's introduction to the 1924 Edition (Brentano 1995, 307). Furthermore, he mentions that Brentano would also say 'ich mache (habe) etwas zum Objekt' (1924, 271n20).

> harking back to Kant, a priori intuitions and concepts were distinguished along with the presentations of experiences. All of Aristotle's achievements were abandoned and that was supposed to be an eminent advance. (Brentano 1995, 287)

That an object of thought does not need to exist is already expressed by Brentano with the term 'objectivity' (*Gegenständlichkeit*), which denotesa property instead of an entity. This property is derived by being directed towards something, that is, through a relation. In the passage about our 'improper use' of language, for instance, he explains the term 'objectivity' by saying that 'all it means is that a mentally active subject is referring to [*bezieht sich auf*] those things' (Brentano 1995, 227 [1925, 158]). Thus, the term 'object' is qualified by the term 'relation'.

Concerning the term 'relation', it is crucial to consider the difference in Brentano's use of the German terms *Beziehung* and *Relation*, which in common use would both translate as the English 'relation'. In the appendix to the 1911 edition, he dedicates a whole chapter to this distinction, starting with the sentence: 'What is characteristic of every mental activity is, as I believe I have shown, the reference [*psychische Beziehung*] to something as an object. In this respect, every mental activity seems to be something relational [*Relatives*]' (Brentano 1995, 211 [1925, 133]). He uses *Beziehung* to describe the intentional relation, and to distinguish this specific relation from the standard logical understanding of relations in philosophy, according to which both relata must exist. By contrast, in an intentional relation, only the fundament must exist; the terminus is not essential. Brentano elaborates on this specific relation as follows:

> What is characteristic of every mental activity is, as I believe I have shown, the reference [Beziehung] to something as an object. In this respect, every mental activity seems to be something relational. And in fact, where Aristotle enumerates the various main classes of his category of ó [relation] he mentions mental reference. But he does not hesitate to call attention to something which differentiates this class from the others. In other relations [Relationen] both terms – both the fundament and the terminus – are real, but here only the first term – the fundament – is real. [...] If someone thinks of something, the one who is thinking must certainly exist, but the object of his thinking need not exist at all. In fact, if he is denying something, the existence of the object is precisely what is excluded whenever his denial is correct. So the only thing which is required by mental reference [psy-

> chische Beziehung] is the person thinking. The terminus of the so-called relation does not need to exist in reality at all. For this reason, one could doubt whether we really are dealing with something relational here, and not, rather, with something somewhat similar to something relational in a certain respect, which might, therefore, better be called "quasi-relational [Relativliches]". (1995, 212–213 [1924, 133–134])

The term 'relation' caused some confusion, and several of his students came up with theories that revolved around the implications of this term and that of the 'immanent object'. Husserl (1901) argues, for instance, that a relationship cannot hold between the mental state of a conscious subject and an immanent object and, on this basis, introduces the concept of 'ideal objects' – objects which transcend the intentional act. On a similar basis, Twardowski (1894) made the distinction between the content and the object of presentations. Moreover, Meinong (1910) developed a 'theory of objects' that investigates intentional objects independently of their existence and claims that such objects have the properties they represent, whether they exist or not. These interpretations and developments led Brentano to offer further clarification regarding his understanding of the notion *psychische Beziehung* on the occasion of the 1911 republication of his work, including of the passage cited above.[14] Notably, he elaborates on the difference between *Beziehung* and *Relation*. Arguably to mark this difference, *Beziehung* is often translated into English as '(mental) reference'.[15] However, this translation might be similarly misleading, because in the case of reference an existential commitment is usually assumed. As Taieb (2017) notes, Brentano explicitly distinguishes between intentionality and reference: the latter obtains only when the object of a mental state exists in reality. In his article, Taieb identifies two major shifts in Brentano's thought: first, a move from his early conception of immanent objects to a quasi-relational account of intentionality (as illustrated in the passage above), since the former would isolate us from access to the external world;[16] and second, a return to a relational understanding of intentionality – one that allows for relations even

14 Cf. Crane's introduction to Brentano (2015, xi) and Kraus's introduction to the 1924 edition (1995, 288ff).

15 Cf. the standard translation of Brentano by Rancurello, Terrell, and McAlister for the 1973 Routledge edition, republished in 1995.

16 Taieb refers to a quote from a letter from Brentano to Kraus, dated 1909: 'it is paradoxical in the extreme to say that what a man promises to marry is an ens rations and that he keeps his word by marrying a real person' (Brentano 1995, 299).

if one of the relata does not exist – while still rejecting the existence of immanent objects.[17]

Evidently, neither the term 'object' nor 'relation' should be understood in their ordinary sense. Brentano does not simply relativise them but rather explains each through the other. On the one hand, the idea that a mental state can 'contain' or 'have' something as its object is explained in terms of its relation to that object. For a thought to have an object means that a subject is mentally relating to something, regardless of whether that object exists. On the other hand, a relation is defined such that 'the one who is thinking must certainly exist, but the object of his thinking need not exist at all'. This specific relationship is further characterized by the way we make something an object of thought: namely, by directing ourselves towards it in a mental act. The mutual explanation of the terms 'object' and 'relation' reveals that Brentano requires both content and directedness to adequately describe the concept of intentionality. One could say that the paraphrases 'content' and 'directedness' reflect the difference between objectual intentionality and directedness, which I pointed out in chapter 2. All the proponents of affective intentionality I have discussed consider a propositional understanding of intentionality too narrow; however, they differ in the importance they assign to the intentional object. We saw that Brentano conceived the notions 'relation' and 'object' as interdependent, so the positions of objectual intentionality and directedness might be reconcilable. They may simply emphasise different aspects of the same phenomenon, depending on their argumentative focus. Instances of intentionality might be most clearly identifiable through their objects; however, this need not diminish the essential role of our directedness towards those objects in making them the focus of our attention. Similarly, the intentional object might, for instance, be conceived as dispositional and only typically present in instances of intentionality, since every direction eventually implies a focus. From the passages in Brentano I have considered, it became clear that any understanding of intentionality must be able to accommodate both aspects. We can conclude that the simultaneous descriptions of intentionality in terms of content and directedness are not problematic; they do not represent two different understandings of intentionality, but rather two complementary aspects of a single phenomenon. This twofold description is what I refer to as the *twofold nature of intentionality*: our mental states have content *by* being directed towards some-

17 For instance, in his later work *Kategorienlehre* (1933, 237–238, 283).

thing, and we relate to something *because* of the importance the intentional object has for us.

4.3 Intentionality as property or capacity

After examining the description of intentionality as content and directedness in Brentano's formulation, I now turn to the descriptions of intentionality as property or capacity. Intentionality without explicit differentiation is sometimes described as a 'property of mental states', sometimes as a 'capacity of the mind'. I believe these paraphrases indicate fundamentally different understandings of intentionality. The distinction between intentionality as a property or a capacity is a further difference I traced in the accounts of Bennett Helm, Sabine Döring, and Jan Slaby. The divergence in their accounts is likely neither the only nor the strongest example of this difference, but it can serve to illustrate and affirm this second distinction.Since they focused on the more specific concept of affective intentionality, we must infer this difference from their general understanding of intentionality, which remains mostly implicit.

4.3.1 Tracing further differences in proponents of affective intentionality

The second differentiation – the paraphrasing of intentionality as a 'property of mental states' or as a 'capacity of the mind' – remains more obscure in the literature, while the implications, as I believe, are much more severe. To make this difference explicit, I will return to the proponents of affective intentionality discussed in chapter 2. We must take a close look at their accounts in order to infer this difference from their general understanding of the term.

Bennett Helm's main concern is the concept of felt evaluation, which enables him to locate evaluation (and, thus, intentionality) in the concept of pleasure and pain. He claims that 'pleasures and pains [...] are essentially intentional and evaluative, a sense of how things are going – whether well or poorly' (Helm 2002, 28). There isn't first a perception that is evaluated in a subsequent step, but the concept of felt evaluation places (affective) evaluation in the very moment of experience. Here, intentionality is formulated as the property of pleasures and pains to be 'about' a state of affairs. He is using the term 'intentional' (next to other adjectives such as cognitive and phenomenal) in order to describe mental states. Thus, his understanding of intentionality is that of a property of mental states, namely, the property of being 'about' or 'of' something. This property can be attributed to some mental states and not others. Intentionality is given particular importance, for 'it is only because such experiences are evaluative that their effects on our attention and motivation are rationally intelligible' (Helm 2009, 253). Not only for Helm but also in the context of emotion theory in general, intentionality has a strong argumentative relevance: it makes affective mental states rationally accessible and assessable. This enables us to understand and evaluate the rational pattern behind emotions and their interrelation with other mental states and behaviours. Döring and Reisenzein claim that due to their intentionality, 'emotions are capable of making other mental states and actions rational, rather than merely causing them' (2009, 201). They further explain that 'in today's terminology, "intentionality" typically amounts to "mental representation": thus the claim that emotional experiences are directed at objects translates into the claim that emotions are representational mental states; states that present objects to the subject in a particular way' (196). Here, intentionality is considered a property of mental states, and the criterion for a mental state to be intentional is to have an object; that is representational or propositional content.

This is the more common understanding of intentionality nowadays. We already saw that, according to Tim Crane, there has been a shift in the use of the term, contrasting the propositional understanding of intentionality with Brentano's notion of intentional in-existence (Crane, 2017). He traced the different use of intentionality in analytic and language philosophy to its introduction by Roderick Chisholm to the English-speaking realm, and argued that, based on Chisholm's interpretation of the term, other (analytical) philosophers came to understand intentionality as having a propositional attitude. I will expand this argument further. I believe that the understanding of intentionality in emotion theory is related to the introduction of the term through cognitive emotion theories and their connection to early language philosophy.Chisholm's understanding also influenced Anthony Kenny, one of the first proponents of cognitive emotion theory. Kenny, for instance, introduced the distinction between the particular and the formal object of emotion, which constitutes a standard of appropriateness for emotions. According to this distinction, emotions can be directed towards any particular object, but they are only fitting if the particular object instantiates the formal object represented by the emotion, that is, what constitutes attraction, danger, and so on. Intentionality is central to this thought, and Kenny explains in the preface to his book *Action, Emotion and Will* (1963) that he decided 'to write [his] thesis on the intentionality of psychological verbs' (2003, vii). In addition to Brentano, there are several references to Chisholm. He acknowledges that 'Brentano's thesis was developed by Chisholm' (Kenny 1963, 138) and speaks of the criterion that 'we have borrowed from the scholastics and from Chisholm' (141). Kenny was familiar with the original text – he uses the edition by Kraus as a reference – and at one point he includes a German citation (1963, 136, 158). Nevertheless, he also read and was influenced by Chisholm's interpretation of intentionality. One of Kenny's primary concerns is to include an account of how sentences containing intentional objects can be analysed in a theory of emotion and action. This corresponds to Chisholm's application of Brentano's concept of intentionality to the analysis of language and his understanding of intentionality as having a propositional attitude. The main concern of cognitive emotion theory was to make emotions and related behaviour rational and assessable. It is easy to determine the suitability of an account of propositional intentionality for this purpose.We saw in the example of Walton how a propositional understanding of intentionality in early cognitive emotion theory was applied to emotional responses to fiction, and what constraints this imposed on his theory.However, the reception of Brentano's concept in the English-speaking realm can provide a broader perspective on the subject. It suggests that this is not the only understanding of intentionality that is justified. One of the points on which adherents of affective intentionality agree is that a propositional conception of intentionality is reductive when it comes to analysing emotions. The intentionality of emotions comes with different correctness conditions, not necessarily involving an existential commitment to its objects. On this basis, I distinguished an *objectual* understanding of intentionality – a mental state 'having an object' – in Döring and Helm from that of having a propositional attitude. However, both propositional and objectual intentionality are understood as a property of mental states.

Slaby also discussed the more specific concept of affective intentionality, but here, we encounter yet a different understanding of intentionality. In his view, the *intentional feeling* discloses the significance of something to us. For him, feelings are 'crucial carriers of world-directed intentionality' (Slaby 2008a, 430), and he explains that if we are affectively relating to something outside our body, the body is 'the vehicle of the intentional feeling' (440–441). He places significance and evaluation in the bodily experience, claiming that 'the felt body is itself that through which we grasp what goes on around us' (436). 'Feeling' in the concept of the *intentional feeling* used by Slaby is derived from a verb. Furthermore, it is not a passive perception but describes an activity: the act of feeling. In this case, intentionality is consequently not understood as a property of mental states but as an activity. It is an orientation *towards*, which does not necessarily need an object nor is necessarily fixed in a state. The intentional feeling is also described by Slaby as 'latent grasping' *and* is explained through the general situatedness of the body in the world.[18] This understanding of intentionality differs significantly from that proposed by Helm and Döring.Slaby's considerations are based on Sartre's situated understanding of existence and Heidegger's interpretation of intentionality as 'behaving-toward' (*Verhaltung*). He gives this understanding a practical turn through the emphasis on the body and describes intentionality as an 'active personal perspective we have in the world', which cannot be reduced to a property of mental states, but is best understood as a mental activity:

> [Intentionality] is not a property of mental states or events, that could be individuated independently from the aspect of such relatedness [Bezogenseins] [...]. Instead, intentionality is understood as an all-encompassing behaving-toward [Sich-zu-etwas-verhalten]. It cannot be adequately defined as a property of something but has to be seen as a fundamental category: [...] intentionality is nothing other than actively being related [tätiges Bezogensein] to the world, oneself, and others. (Slaby 2008b, 111, my translation)

He insists on the notionof 'being active' (*Tätigsein*), and activity here is understood in a broad sense, encompassing seemingly passive (mental) behaviour like dwelling on or contemplating something.[19] There are two dimensions of activity in his view: the purposeful activity, on the one hand, and the necessary activity on the other.[20] The latter sense refers to finding oneself in a situation and reacting to circumstances because they are relevant to us in a certain re-

---

18 A chapter in the coedited volume *Affektive Intentionalität* (2011), called 'Affektive Intentionalität – Hintergrundgefühle, Möglichkeitsräume, Handlungsorientierung', which I referred to earlier.

19 '...wenn man sich vergegenwärtigt, wie umfassend der Bereich der Tätigkeiten tatsächlich ist: selbst so passivisch anmutende Verhaltungen wie ein *Verweilen bei* oder *Betrachten von* etwas sind Tätigkeiten im vollsten Sinne des Wortes' (Slaby 2008b, 112n7).

20 In this sense of activity we are always active, we always behave-towards something, since we always find ourselves in certain circumstances/'the world'. This corresponds to Slaby's position of agreeing with Brentano concerning the claim that all mental states are intentional.

spect. This makes the necessary activity intentional (in the technical sense): it is a behaving-towards something that is experienced as relevant.

Slaby claims that affective intentionality is a different kind of intentionality. However, the difference seems to lie rather in the different general understanding he brings to it. Döring and Helm understand intentionality as a *property* of mental states, while Slaby understands it as a mental *activity*. These are fundamentally different understandings of intentionality and to use the same term for both phenomena amounts to an equivocation. Hence, it is crucial to differentiate between the two uses of the term.

4.3.2 Intentionality as capacity to have intentional mental states

There is a variant of understanding the descriptions 'property of mental states' and 'capacity of the mind' that could bridge this difference, namely, to understand intentionality as the 'capacity of the mind to have intentional states'. This distinction could be a means of avoiding the problem of equivocation. If we continue with the examples of affective intentionality, we should be able to reformulate Helm and Döring's as well as Slaby's understanding of intentionality as the 'capacity of the mind to have intentional states'. Can both positions be reformulated in this way?

Helm and Döring's general understanding of intentionality is that of a property of mental states. This is not to be understood in Brentano's sense that intentionality is a property belonging to all mental states – a mark of the mental – but as a property some have and others lack. This interpretation is compatible with the formulation that intentionality is the capacity of the mind to have intentional states. The main assumption – that intentionality is a property – is preserved. The mental phenomenon denoted is a mental state and the property of being intentional is attributed to it. 'Capacity' describes a potentiality here: the potential of having mental states, which have the property of being intentional. Thus, it would be possible to reformulate their accounts in terms of the description of intentionality as a the 'capacity of the mind to have intentional states'.

Slaby's understanding of intentionality seems, at first glance, also to be compatible with this description. Upon closer examination, however, this seeming compatibility turns out to be superficial. This is due to the notion of 'capacity'. 'Capacity' merely names a potential or aptness for something; it does not specify a class of mental phenomena. The capacity to have intentional mental states differs from the capacity to be intentionally directed, as the latter refers to the capacity for an activity. We saw that, for Slaby, intentionality is essentially a mental activity – a behaving towards. It does not necessarily have an object and is not necessarily fixed in a propositional state. Nevertheless, it has a 'propositional disposition' (*begriffliche Verfasstheit*). This means that, if necessary, it can be cast as propositional content.[21] From this perspective, intentionality can be instantiated as a property of mental states but is not limited to it. As a property of mental states, intentionality for Slaby is 'an abstraction – fundamentally and primarily intentional is the whole active being [*das tätige Sein*] of a person in the world' (2008b, 112). These explanations clarify in which sense Slaby primarily understands intentionality as an activity of the mind (through the vehicle of the body). His understanding is fundamentally different from that of Helm and Döring: he understands intentionality as a mental activity that can be, but does not have to be, instantiated in a propositional form. Thus, the term 'activity' is more appropriate than 'capacity' for describing this understanding.

Slaby's general understanding of intentionality is fundamentally different from Helm's and Döring's, and rephrasing intentionality as the 'capacity of the mind to have intentional states' will not resolve the differences we encounter between them. Not all accounts can be rephrased in this way, as doing so would fail to do justice to the notion of 'activity'. The reformulation merely shifts the problem to a different level. We would still need to distinguish between the capacity for having a property and the capacity to engage in an activity. It becomes evident that there is a fundamentally different understanding of intentionality at play. For a clearer delimitation it further makes sense to speak of intentionality as an activity of the mind, instead of a capacity. In the following I am going to use the differentiation intentionality as 'property of mental states' or as 'activity of the mind'.

The understandings of intentionality as a property of mental states and as an activity of the mind are fundamentally different, and, if not differentiated, using the same term for both can lead to category mistakes.There is reason to believe that the differences encountered in the accounts of Helm, Döring, and Slaby concern not only the restricted domain of affective intentionality but exemplify a larger ambiguity in the use of the term.

---

21 Slaby refers to Tugendhats interpretation of Heidegger (cf. Slaby 2008b, 114n9) and further discusses a broad understanding of propositionality in chapter 10.

4.3.3 Mental states and mental acts

We have seen that it is necessary to distinguish intentionality as a property of mental states and as an activity of the mind. However, is this not the same difference that was already drawn between mental states and mental acts?

In the philosophy of mind this distinction is sometimes made, and it is used in slightly different ways. Mental states are relatively stable and enduring, they are often understood as dispositions, meaning they can exist without being consciously entertained at any particular moment. Beliefs, desires, fears, and intentions are examples of such states. These states are not always conscious, but they are understood as representationally structured – that is, they have intentional content. In contrast, mental acts are temporal occurrences – episodes in which the mind actively engages with content. Judging, deciding, imagining, and remembering are all acts in this sense. Mental acts are conscious, and intentional as they are directed towards objects, propositions, or states of affairs.

In this text I have not distinguished between mental states and mental acts in this sense, but I have referred to both as mental states. I follow Searle's terminology in this. He often refers to both as mental states; however, he emphasizes the distinction between dispositional and occurrent. For him, dispositional states typically underlie or generate speech and action, while occurrent states are their conscious or performative expressions. He does not make a clear distinction between beliefs and desires as states or acts, but rather between their direction of fit: beliefs aim to match the world (*mind-to-world*), while desires aim to shape the world (*world-to-mind*). In Searle's view, belief can be both a state and an act: the former is dispositional while the latter is occurrent, but both have the same intentional structure. They both have a propositional content and a psychological mode. And independently of whether a belief is dispositional or occurrent, it can play a causal role in a system.

The difference between mental states and mental acts cannot be equated with that between intentionality as property and activity. In particular, we have to be careful not to confuse mental activity with mental acts. A mental act is the active temporal occurrence of a thought, and is distinguished from a disposition which is passive, not necessarily conscious, but more enduring. The understanding of intentionality as a mental activity refers to something different. It denotes the activity of being (mentally) directed towards something. It is durational, does not necessarily have an object, and is not necessarily fixed in a propositional state. Nevertheless, it hasa 'propositional disposition'. This means that, if necessary, it can be cast into propositional content; that is, we can describe what is relevant to us in a certain situation, how we relate to it, and (often also) why.[22] From this perspective,intentionality is understood as a mental activity, one which can but does not have to be instantiated in a propositional form. Thus, mental acts cannot be equated with this sense of mental activity.

If we return to the descriptions 'property' and 'capacity', we can understand them according to Searle's notions in the following sense: intentionality as a property of mental states is occurrent and intentionality as capacity of the mind is dispositional. This would be expressing that there can be both occurrent and dispositional instances of intentionality, which is not problematic. In fact, I suggest that when the paraphrase 'capacity' is used in lexica and introductory texts, it usually refers to this dispositional sense. Although it is often considered too narrow, the standard understanding of intentionality is still that of propositional intentionality. It is what the contemporary shorthands 'aboutness' or 'ofness' suggest: that some state S is intentional if there is an 'answer to such questions as: What is S about? What is S of? What is it an S that?' (Searle 1983, 2). This narrow understanding is the cause of many difficulties that we encounter outside the more particular discourse on intentionality, where the term is mostly used in this way. In this view, intentionality is understood as a property of mental states which is sometimes occurrent, that is, instantiated, and other times dispositional as capacity. To use the paraphrases 'property' and 'capacity' in this sense would be unproblematic. The discrepancy I have found in the general understanding of intentionality between proponents of affective intentionality indicates, as we saw, a more fundamental, and problematic, difference. To understand intentionality as a mental activity that is durational and can, but does not have to, be instantiated in propositional form differs from understanding it as a property (of a mental act) that is either occurrent or dispositional. The notions of property and activity designate different kinds of mental phenomena, and using the same term for both is a case of equivocation and can lead to category mistakes.

4.3.4 The notion of mental activity in Brentano

For Brentano, all mental phenomena – whether beliefs, judgments, desires, or presentations (*Vorstellungen*) – are characterised by their intentional direct-

22 Cf. Slaby (2008b, 114n9).

edness and are consistently described by him as forms of mental activity. He speaks, for instance, of '(psychische) Akte', 'psychische Tätigkeit', 'Denkakte', 'Empfindungs- und Vorstellungsakte', and 'Seelentätigkeit'.[23] Indeed, the emphasis on mental activity is significant in Brentano's work. He uses the phrase 'the mentally active' (*ein psychisch Tätiges*) both as a characterisation and, in some cases, even as a synonym for a subject. For example, in the following sentence:

> Because reference to things [sich auf Dinge zu beziehen] is a distinctive characteristic of someone who is mentally active [des psychisch Tätigen], we have been led to talk about objects having being or subsisting in the mentally active subject [dem psychisch Tätigen]. (Brentano 1995, 227 [1925, 158])

In this sense, intentionality is understood as a mental activity. The English translation 'mental reference' obscures this notion of the mentally active, which Brentano insists upon. Notably, intentionality is described as the *activity* of mental reference (*das psychisch sich Beziehende*) and is used by him to determinethe mental realm. 'Alles psychisch sich Beziehende bezieht sich auf Dinge' (1925, 158) is, for instance, translated as 'All mental references refer to things' (1995, 227). Similarly, the term *psychische Beziehung* was translated as 'reference'. However, we saw earlier that he clearly distinguishes intentionality from reference (Taieb 2017).[24] Moreover, the translation of the term *Beziehung* as 'reference' favours a linguistic reading of intentionality and diminishes the importance of the notion of activity. Thus, it suggests a reductive sense of intentionality. The German *Beziehung* can be translated as 'relation' or as 'reference'; both, however, involve an existential commitment as they are used in the philosophical literature today. This implication is not intended, and verbs such as 'attending to' or 'relating to' would perhaps better describe the mental activity Brentano had in mind.

Furthermore, Brentano emphasises the process of mental activity by distinguishing it from the object of the act, for instance in the distinction between hearing and what is heard. He points to a source of confusion:

23 A non-exhaustive list; all page numbers refer to Brentano (1924): '(psychische) Akte' (142, 179, 182, 187, 201, 203, 218), 'psychische Tätigkeit' (236, 246, 250), 'Denkakte' (245), 'Empfindungs- und Vorstellungsakte' (250), and 'Seelentätigkeit' (144, 174, 175, 193, 228, 237).

24 Cf. section 4.2.3.

> Of course these terms, as such, can also be used in a different sense, in which "presentation" means not only the act of presenting but also the object which is presented, and "sensation" means not only the act of sensing, but also the object which is sensed. In our case, however, we are dealing with mental activities. (Brentano 1995, 135)

He remarks in a footnote to this passage that 'many psychologists and epistemologists are still guilty of this confusion today' (ibid); and, in the next paragraph, speaks of 'an obvious fallacy of equivocation, [when] the word "presentation" [is] used first in the sense of the object presented, and then in the sense of the act of presenting it' (1995, 136). When we perceive something, we are not simply 'having' a mental representation; rather, we engage in the *activity of perceiving* that object. Similarly, judging is an active evaluation of a proposition, loving or hating involves a prolonged engagement with an object or person, and desiring entails a sustained orientation towards what is sought.

As we saw in the previous chapter, we must be careful not to conflate mental activity with mental acts. The contemporary use of the term 'mental act' refers to the active, temporal occurrence of a thought as distinct from a disposition, which is passive, and not necessarily conscious. Furthermore, a mental act often instantiates a propositional attitude. Brentano's notion of mental activity does not conform to this distinction.It should not be confused with the contemporary usage of'mental act', as he also includes non-propositional and temporally extended phenomena, such as perceptions. Furthermore, although he acknowledges the 'fact that we can possess a store of acquired knowledge as a disposition towards certain acts of thinking' (1995, 79), he is not concerned with the distinction between dispositional states and mental acts. He instead engages in a sustained argument against theorising about unconscious mental states: we can neither directly experience them (which would be absurd)[25] nor inferentially presuppose them.[26] While he acknowledges that dispositional states do exist, he has already restricted his analysis to occurrent mental phe-

25 'Denn, dass keine unbewußten Vorstellungen im Bereiche unserer Erfahrung vorkommen, ist selbstverständlich und notwendig der Fall, auch wenn solche in reicher Anzahl in uns vorhanden sein sollten; wären sie ja sonst nicht unbewußt.' (Brentano 1924, 147).

26 Over the course of more than forty pages, he argues against four possible accounts of such an inferentially grounded existence: as cause, as effect, as grounded in the correlation between the strength of the phenomenon and its consciousness, and as grounded in the argument from infinite regress.

nomena (*wirkliche Zustände*).[27] Brentano uses the notion of 'mental activity' in a broad and quite literal sense to describe a fundamental and constitutive aspect of the mind. Furthermore, the notion of mental activity emphasises that mental phenomena are actively engaged with their objects, rather than merely being 'about' them.

Thus, the understanding of intentionality as a mental activity is already present in Brentano. However, this is not the only sense in which he uses the term. The famous passage, 'Every mental phenomenon is characterized by what the Scholastics of the Middle Ages called the intentional (or mental) inexistence of an object' (1995, 168), suggests that mental states *have* certain features: they include an object or are directed towards something. He continues to specify: 'In presentation something is presented, in judgement something is affirmed or denied, in love loved, in hate hated, in desire desired and so on.' (Brentano 1995, 68). This resembles propositional attitudes and indicates that intentionality is treated as a *property* of those mental states – something they possess, or that can be predicated of them. Indeed, both understandings of intentionality, as a property of mental states and as an activity of the mind, seem interrelated. On the one hand, intentionality as a property suggests that mental states (whether occurrent or dispositional) have the property of being 'about' something – that they *have* directedness. On the other hand, conceptualising intentionality as a mental activity emphasises a dynamic process whereby the mind *actively directs* itself towards an object. Both understandings of intentionality – as a property of mental states and as an activity of the mind – are present in Brentano; however, he does not seem to explicitly differentiate between them.

One way in which Brentano might (implicitly) be addressing this difference is through the so-called *dual relation of the mental* (*Doppelbeziehung des Psychischen*): he held that every mental phenomenon has an object *and* is conscious. 'Every mental activity is accompanied by a consciousness which refers to it' (1995, 107).[28] The latter is also referred to as 'inner consciousness', in the sense that its object is the mental activity itself, while the former has something other

---

27 'Als eigentlichen Gegenstand der Psychologie werden wir nur die psychischen Phänomene in dem Sinn von wirklichen Zuständen anzusehen haben.' (Brentano 1924, 140).

28 Partly my translation: although Brentano uses the term *psychischer Akt* here, we should not confuse mental activity with mental acts, as shown in the previous chapter. Thus, I rendered 'act' as 'activity'.

than itself as an object (be this physical or abstract). Textor (2013) formulates this thesis as follows: 'every mental phenomenon M is primarily directed upon an object other than M and secondarily (concomitantly) upon M itself in a way that yields knowledge of M'. My aim here is not to discuss the dual relation thesis; it faces several objections and Brentano's terminology must be carefully unpacked, since his philosophical framework differs from the one we use nowadays. I mention it as an indication that he might have been aware of the difference between intentionality as activity and as property. To better understand his claim, a short note on his terminology: 'inner consciousness' is immediate knowledge of mental phenomena in the sense of their acknowledgement. It does not necessarily take a propositional form but is more generally conceived as presentation (*Vorstellung*). Furthermore, this immediate and noninferential acknowledgement of the mental activity happens 'en parergo':[29]

> In the same mental phenomenon in which the sound is present to our minds we simultaneously apprehend the mental phenomenon itself. What is more, we apprehend it in accordance with its dual nature insofar as it has the sound as content within it, and insofar as it has itself as content at the same time.* [...] The act of hearing appears to be directed toward sound in the most proper sense of the term, and because of this it seems to apprehend itself incidentally and as something additional. (Brentano 1995, 98)

Brentano begins with examples from perception because he needs non-inferential instances which only involve presentation (*Vorstellung*) – the basic form of an intentional mental phenomenon in his conception – and on which other phenomena, like judgement, belief, love/hate, are then built.[30] In the above quote, it becomes clear that Brentano's focus lies in explaining that the activity of hearing and the acknowledgment of that activity are not two temporary separate phenomena but happen simultaneously. In fact, he maintains that the two relations are aspects of a single mental phenomenon with a 'dual nature'. Much of his work on this *dual relation thesis* concerns the unity of consciousness,

---

29 A notion from Brentano's translation of Aristotle's *Metaphysics* 12.9 (Brentano 1995, 102).

30 Cf. chapter 3, *Further Considerations Regarding Inner Consciousness* (1995, 107ff), where he is discussing the complexity of this 'accompanying consciousness': that is, how presentation, judgement, belief, and affect relate to each other in the mental act.

which in turn bears on how an object can be immanent in a mental act. However, what makes this challenging is that these two aspects are quite different. First, they have different objects – namely, the sound and the mental activity of hearing it. He even draws a distinction between a primary and a secondary object in this respect.[31] But more importantly for our purposes, they also seem to differ in nature. Brentano distinguishes between the act of hearing and the acknowledgement of what is heard. At some point, he writes:

> The presentation [Vorstellung] of the sound is connected with the presentation of the presentation of the sound in such a peculiarly intimate way that its very existence constitutes an intrinsic prerequisite for the existence of this presentation. (Brentano 1995, 98)

According to Textor's interpretation of this passage, what constitutes an 'intrinsic prerequisite for the existence' of the other mental act (and what Textor himself translates as 'contribution to the being' of the other mental act) is 'the fact that the acknowledged mental act individuates its acknowledgment. The acknowledgment is fully specified if one specifies which mental act is acknowledged' (2013, 476). He argues that this makes acknowledgment a 'primitive factive mental act'. For our purposes, it is particularly important that Textor describes the aspect of acknowledgment as 'individuating' the content of the mental activity. Such individuation involves *attributing* intentionality to that phenomenon – that is, treating intentionality as a property of mental states. In this sense, the understanding of intentionality as both activity and property is implicit in Brentano's dual relation thesis: as the mental activity of *directedness*, and the (simultaneously occurring) acknowledgement of the very same act *as being directed*.

Brentano was more concerned with showing that these two aspects are part of the same phenomenon and occur simultaneously, rather than with their differentiation. He is particularly interested in reconciling the notions of content and directedness – which I treated in the last chapter and described as the twofold nature of intentionality – and thereby explain the notion of 'intentional inexistence' (mentale Einwohnung). By tying the intentional content inseparably to the mental act itself, he avoids, for instance, treating the object of the intentional act as an independent entity. We have seen that, for him, intentionality is essentially a mental activity, in which 'having a content' is an

31 Cf. Brentano (1995, 98).

integral part of this activity – something that emerges *en parergo*. In this sense, it can be said that the understanding of intentionality as a property of mental states and as an activity of the mind is already differentiated in Brentano's dual relation thesis. However, this does not acknowledge the conceptual problems that may arise from conflating the understandings of intentionality as property and as activity. I believe that it is crucial to distinguish these aspects of intentionality.

Although they may denote different aspects of the same mental phenomenon, the equivocal use of the term intentionality as property or activity can lead to category mistakes and conceptual confusion. This becomes evident in the way intentionality is conceptualised and used in different contemporary discourses – namely, that it is often reduced to *one* of these aspects. Treating intentionality (solely) as a property can, for instance, implies that intentional mental states must necessarily have an object that exists. In such cases, the intentional object is taken to be the marker of intentionality, and the intentional relation is thought to obtain only if this object has a real referent. This was evident in Walton's theory, where we saw how a propositional understanding of intentionality constrained his account. Furthermore, if we also ascribe intentional properties to artefacts and do not clearly distinguish between the understanding of intentionality as property and as activity, this can lead to the assumption that intended content can be securely derived from those properties – that they somehow put us 'in touch' with that content. This amounts to conflating intentional properties with (intrinsic) intentionality. However, as we have seen, only function derivation, not content derivation, reliably obtains when attributing intentional properties to artefacts. Moreover, clearly differentiating the notions of activity and property helps avoid the category mistake of attributing mental states to nonconscious items or systems.

Thus, distinguishing intentionality as a mental *activity* from intentionality as a *property* of mental states can provide a framework for addressing issues where these notions have been reduced or conflated. This distinction might also help further clarify Brentano's complex and sometimes ambiguous concept of intentionality and encourage a deeper understanding of how mental phenomena relate to their objects.

4.4 Conclusion

I set out to use the paradox of fiction as a tool to analyse the concept of intentionality. What emerged were divergent – and potentially incompatible – conceptions of the term, which often remain implicit. At the end of chapter 2, I concluded that it is not necessary to posit the intentionality of emotions as a distinct kind, but rather to differentiate between the formal objects of emotions and those of beliefs and desires, which implies that these states have different correctness conditions and entail different existential commitments. However, it has become apparent that more fundamental differences are at play: differences in the 'general' understanding of intentionality. To approach these often-implicit understandings, and to navigate the vast domain of intentionality, I examined the various paraphrases commonly used to describe the term: namely, intentionality as *content* and *directedness*, and as a *property of mental states* or a *capacity (or activity) of the mind*.

Regarding the first pair, I concluded that Brentano relies on both notions for his concept of intentionality. Notably, he does not merely relativise them but explains each in terms of the other: our mental states have content *by* being directed towards something, and we are directed towards something *because* of the importance the intentional object has for us. The distinction between intentionality as a *property* of mental states and as an *activity* of the mind, however, proves more problematic. These represent fundamentally different understandings of intentionality, and using the same term for both is a case of equivocation and can result in category mistakes and conceptual confusion. Although I examined this distinction through a specific example – namely, differing views among proponents of affective intentionality – it can be assumed that this is not an isolated case but symptomatic of a broader conceptual issue.

The question now arises: how did we arrive at these different understandings of intentionality – and, more importantly, which of them is more justified? Some points concerning the development of the term in different traditions have already been mentioned. The propositional understanding of intentionality in early cognitive emotion theory, for instance, must be seen in light of how the term was introduced to the English-speaking realm, and the connection between cognitive emotion theories and early language philosophy. According to Crane, there has been misinterpretation of Brentano's concept in the 20th century, which can be traced back to how Brentano was introduced to the English-speaking world by Chisholm. On this basis, other analytic philosophers came to understand intentionality as the 'having of a propositional attitude',while Brentano's original term aimed at providing a criterion for distinguishing between physical and mental phenomena, and was not primarily concerned with the structures of language. This aligns with my findings concerning the introduction of the term into cognitive emotion theory by Kenny, particularly his conflation of 'intentionality' and 'intensionality' in this context. Surely, contrasting the analytic understanding of intentionality with its lineage in the continental tradition could be of interest. However, this is not the place for such an endeavour. Rather, I showed that both understandings – as property and as activity – are already present in Brentano's writings. Nevertheless, he did not explicitly differentiate between them. Brentano essentially conceives of intentionality as a mental activity, while at the same time describing mental states as having the property of containing the object to which they are directed. I suggested that the differentiation between intentionality as activity and as property is most evident in his so-called dual relation thesis (the claim that 'every mental activity is accompanied by a consciousness which refers to it'): namely, the simultaneous occurrence of the mental activity of *directedness* and the acknowledgement of the very same act *as being directed*. Brentano is primarily interested in reconciling these two aspects of intentionality to maintain the unity of consciousness. He distinguishes between intentionality as a property and an activity; however, he does not anticipate the problems that may arise from conflating these notions. I hold that the first step must be to make this distinction explicit. Although they may denote different aspects of the same mental phenomenon, the equivocal use of the term intentionality to refer either to a property or to an activity can lead to category mistakes and conceptual confusion. A clear distinction can provide a framework for addressing cases in which these notions have been conflated, as well as invite a deeper appreciation of how mental phenomena relate to their objects.

While the twofold description of intentionality as *content* and *directedness* can be used synonymously and reflects the twofold nature of intentionality, the descriptions of intentionality as a *property* of mental states or as an *activity* of the mind must be differentiated. However, this differentiation need not be exclusive; rather, it may be productive.

5. Results, requirements, and *relatedness*

> So many different kinds of stones,
> the sober and the uncanny, lying side by side in the red dirt.
> To stop and imagine the life of each one!
> Now they were sailing through the air
> from a happy human arm,
> what a fate. Geryon hurried on.
> *Anne Carson, Autobiography of Red*

In my study I have examined various explicit and implicit understandings and uses of the term 'intentionality' through the two perspectives offered by the paradox of fiction. These perspectives were initially formulated as the questions (A)', whether there is a difference between the intentionality of thought and that of emotion, and (B)', whether there is a difference in intentionality when directed at a factual or a fictional state of affairs. Intentionality lies at the core of the paradox: our emotional responses are both experienced and formulated as directed *towards* the fictional characters and situations. Here, the paradox is not treated as a problem to be solved, but as a tool to examine the concepts involved. Just as the concept of emotion has undergone critical reevaluation in the course of this debate, I argue that challenging the concept of intentionality is equally fruitful – not only for clarifying Ref but also for refining our understanding of intentionality.

To explore these questions, I focussed on two prominent examples from the discourse on the paradox of fiction – namely, Walton's *three-men example* and Radford's *tragic-sister example* – and discussed two relevant concepts of intentionality, accordingly: affective intentionality and derived intentionality. A close examination of the largely implicit understanding of intentionality underlying these examples shows that it has often been either neglected or treated reductively. The concept of intentionality has not yet been questioned directly in the context of the paradox and is rarely mentioned at all. I addressed the two perspectives (A)' and (B)' in the prospect that their intersection could offer valuable insights into the nature of intentionality. Approaching intentionality from these two angles not only provided orientation within the vast field of the debate, but also revealed inconsistencies in the understanding of it, and conjointly offered a more comprehensive view of the problem.

In examining the concepts of affective and derived intentionality, I have paid particular attention, throughout the study, to the 'general' understanding of intentionality that underlies the different authors' accounts. The observed discrepancies, particularly between advocates of affective intentionality, led me to examine more closely the different paraphrases typically used to describe intentionality in chapter 4. While paraphrasing the concept in terms of content and directedness reflects what I call the *twofold nature of intentionality*, the descriptions of intentionality as a property of mental states or as an activity of the mind must be differentiated. In this final chapter, I draw together the results of my analysis to outline a set of requirements that a theory of intentionality should meet. Furthermore, I introduce a new paraphrase for intentionality that emerged during my study: the notion of relatedness.

5.1 Results

In chapter 2, I examined question (A)': whether the intentionality of thought differs from the intentionality of emotion. There is a common intuition that the difference between whether the objects of our emotions or those of our thoughts exist is a significant one. I pointed out two primary aspects in the literature connected to this question: first, that emotions typically involve tendencies or dispositions to act, and, second, that intentionality is not necessarily attributed to all emotions.

Emotional responses to fiction are often characterised as having reduced behavioural output. I analysed one of the most prominent examples of this claim: Charles fearing the slime, from Kendall Walton. It has been shown (cf. Werner, 2020) that reduced behavioural output cannot function as a distinction criterion between emotional responses to fictional and factual situations, as Walton suggests. I proceeded to examine the role intentionality plays in Walton's theory, and to clarify his implicit understanding of intentionality. Walton adopts a propositional view: intentionality is understood as having a propositional attitude towards an object. This is tied to the assumption that genuine emotions require the existence of their objects. Walton's position reflects the dominant understanding of intentionality in analytic philosophy at the time, especially within early cognitive theories of emotion. I argue that this propositional model of intentionality contributed to the formulation of proposition (2) of the paradox of fiction: that in order to experience emotional responses towards an object, we must believe in its existence. My analysis showed that intentionality is one of the main concerns and constraints of Walton's theory, even if it is not explicitly acknowledged as such.

Walton – and arguably the majority of theorists in the debate – assume that there is no essential difference between the intentionality of emotions and that of thoughts. The propositional understanding of intentionality in emotions relates to a narrow cognitivist view: here, intentionality is treated as unquestioned in beliefs and desires, and only selectively attributed to emotions. In this understanding, not all emotions are considered intentional, since not all possess a determinable object they are 'about' or 'of', as is the case, for instance, with moods. This places primary emphasis on the intentional object, which is often treated as a marker of intentionality. Ref complicate this picture, as the object of emotion in these cases is fictional. In response, theorists have developed various strategies: reducing Ref to reflexive affective responses (Choi), proposing a substitute object (Lamarque), or relocating the emotion into the fictional world to avoid 'cross-world-saving or handshaking' (Walton 1990, 195).

In the second part of the chapter, I examined the role of intentionality in emotion theory more broadly. Even if we reject the view that a genuine emotion requires the existence of its object, there are good reasons to attribute intentionality to emotions, including to Ref. Accounting for the intentionality of emotions was and is one of the main challenges for emotion theory. It plays a role on three levels: it provides the emotional experience with its 'about', making emotions amenable to rational assessment; it figures in definitions that differentiate intentional emotions from nonintentional feelings; and, historically, it helped integrate emotion into broader philosophical debates, particularly in the context of the *cognitivist turn* in emotion theory.

I then turned to the concept of affective intentionality. Its proponents generally claim that the intentionality of emotions must be treated differently from that of beliefs and desires. At first glance, this would seem to answer (A)' in the affirmative: the intentionality of emotions is different from the intentionality of thoughts. However, my analysis revealed important divergences among proponents of affective intentionality, both in their accounts of affective intentionality and in their general understanding of intentionality. Not all claim that the intentionality of emotions is an intentionality sui generis. Some authors, such as Döring and Helm, argue instead that intentionality 'works' differently in emotions than it does in beliefs and desires.On this view, the intentionality of emotions is associated with different correctness conditions and does not necessarily involve existential commitment. Intentionality operates differently in emotions, not because it is of a fundamentally different kind, but because it has different formal objects: while the formal object of a proposition is 'truth', the formal object of fear, for instance, would be 'danger'. They adopt what I call an *objectual* understanding of intentionality, that is, the 'having of an object'. This does not require propositional content – a content expressible in a sentence that can be true or false – but only an object the mental state is about. For Döring and Helm, the intentional object remains the primary marker for intentionality.By contrast, Jan Slaby argues that affective intentionality constitutes a different *kind* of intentionality. He understands affective intentionality as a different relation we have to the world and oneself (*Welt- und Selbstbezug*); one that is situated and embodied. While Slaby's is a sui generis theory of intentionality, Döring's and Helm's views are rather sui generis theories of emotion.

Despite these differences, all proponents of affective intentionality agree on the so-called *unification view*: that the phenomenality of emotions and their intentionality are inseparable in emotional experience. I examined what this entails more closely, identifying what I call the aspect of *inferential encapsulation*. Inferential encapsulation (ie) is the informational isolation of certain mental states from other beliefs and knowledge that we hold. I proceeded to apply the unification view to Walton's examples, which proved useful. The inseparability of phenomenality and intentionality, and the related aspect of inferential encapsulation help explain the phenomenon of 'conflict without contradiction' characteristic of certain emotions – namely, why some emotions can persist despite conflicting beliefs or the subject's better knowledge. Inferential encapsulation can explain how emotions can isolate, or 'shortcut', rational inference in the moment of affect. This, in turn, helps account for the reduced behavioural output typical of some emotional responses to fiction. The unification view, and particularly its aspect of inferential encapsulation, thus offers considerable advantages in explaining Ref: it allows us not only to refute the charge of doxastic contradiction but also to dismiss the accusation of practical irrationality.

Furthermore, the phrasing of (A)' needs to be reconsidered, since the distinction between thoughts and emotions cannot be clearly drawn. Most of our mental states are intricately intertwined conglomerates of cognitive, affective, and embodied aspects. The examples typically used in the different theories of intentionality are already separated from the outset: in general discussions of intentionality, emotional responses are often excluded, as they seem to represent a special case. Conversely, theorists who focus on the intentionality of emotions tend to consider only emotional responses, with their observations and conclusions often differentiated from – or even opposed to – the intentionality of beliefs and desires. Yet our responses cannot be neatly divided into thoughts and emotions; they are interrelated and overlap in our experience. Of course, it can be useful to distinguish different instances of intentionality for the purpose of theoretical analysis. However, the results of my study show that it is not the difference between thoughts and emotions but rather that between emotions and beliefs/judgements that deserves particular attention. A belief or judgement requires an existential commitment to its intentional object, due to its truth condition. However, different correctness conditions apply to emotions: they have different formal objects and do not necessarily involve an existential commitment. Beliefs are a specific kind of thought: thoughts with mind-to-world direction of fit. As such, beliefs and judgements necessarily imply an existential commitment, while emotions don't. Regarding (A)', I concluded that we do not need to posit a different kind of intentionality for emotions, but rather recognize that emotions have different correctness conditions than beliefs and judgements.

Finally, I traced a different implicit 'general' understanding of intentionality in Helm, Döring, and Slaby. Although all are proponents of affective intentionality and distance themselves from a propositional view, their 'general' understanding of intentionality diverges. Given their emphasis on the object of emotion, I have described Helm's and Döring's accounts as 'objectual intentionality'. For Slaby, on the other hand, intentionality does not necessarily require an object but characterises our situated and embodied relation to the world and is therefore better described as 'directedness'. There is, however, another, more subtle difference that can be observed in comparing these accounts: that of understanding intentionality as a property or a capacity, which I further analysed in chapter 4.

In Chapter 3, I turned to question (B)': whether intentionality is different when directed towards a fictional or factual state of affairs. Although the concept of intentionality was originally developed to accommodate the fact that mental states are directed towards things that may not exist, thought about the nonexistent became one of the 'problems of intentionality'. The paradox of fiction can be seen as one iteration of this problem, since fiction-directed intentionality lies at the heart of the debate. Reconsidering Radford's tragic sister example, I argued that the decrease in empathy it describes is not due to the fictionality of the events, but to the fact that a factual account *turns out* to be fictional – that is, to a change in the status of the account. Such a change constitutes a major shift, as the conditions under which we view the account are now different. When something we assumed to be real turns out to be made up, we are taken aback and must reconsider how to relate to it. Our disengagement is thus caused by a change in the conditions under which we register the account, not the fictionality or factuality of the account itself. One way in which such conditions are indicated to us is the form in which the account is presented. The *presentational aspect* has been largely neglected in discussions of the paradox. It concerns not only *how* something is presented, but the fact that it is: when we are dealing with fictions, we are always dealing with representations. This point is also emphasized by Frank Zipfel, who reformulates the paradox of fiction as the paradox of mimesis. Based on these observations, I ask what implications this has for intentionality and reformulate question (B)' as:

B" Is there a difference in intentionality depending on whether it is mediated or immediately directed at a state of affairs?

Since fictional characters and situations are necessarily encountered through some form of presentation, this led me to inquire about the intentionality of artefacts. Sentences, images, and other artefacts seem to have the property of being directed at, of being 'about' something, in the same way that mental states are. However, intentionality is typically attributed only to mental states, not to physical items or events. This raises the question: Do artefacts also exhibit a form of intentionality, or rather, does it make sense to analyse them in these terms? To explore this question, I turned to Searle's concept of derived intentionality. His theory aims to account for the intentionality of speech acts, and he proposes a particular relation between the intentionality of mind and the intentionality of speech acts: that of derivation. According to Searle, the intentionality of language and artefacts is derived from the intrinsic intentionality of the mind. I presented his theory and evaluated its advantages and disadvantages for explaining Ref in particular, and questions around the intentionality of artefacts in general.

Searle's concept turns out to be useful for addressing the presentational aspect of artefacts. It substantiates the claim that aspects of presentation cannot be excluded from the debate, since artefacts necessarily involve the performance, production, use, or presentation of some physical entity. Furthermore, it is the 'form of presentation' (the equivalent of the intentional mode) that individuates a speech act – regardless of whether it represents a fictional or a factual state of affairs. It helps distinguish, for instance, whether an utterance is a claim, a wish, or a demand. Regarding reduced behavioural output, the concept of derived intentionality further helps clarify why our responses differ when engaging with representations. However, since the term 'reduced behaviour' places too much emphasis on immediate (re)action, I proposed instead to speak of 'indirect behaviour'. In Ref, we are not confronted with something directly but through mediation, and our behaviour is not necessarily reduced.

Two main problems arise with Searle's concept of derived intentionality: its conditionality on the author's intention, and the implications of the adjective 'derived'. Searle's model presupposes that 'meaning intention' is necessary for intentionality, but I showed that this does not hold in all cases. In cases of unintentional intentionality, such as pareidolia, neither the author's nor the receiver's intention is required. Here, we are confronted with accidental agglomerations of signs or marks; however, we are still able to understand a text or recognise an image. The link Searle establishes between intention and intentionality is too rigid and may also lead to confusion with normative questions.Intentionality and meaning should not be conflated, and it is important to distinguish the question of intentionality from the debate around the author's intention in literary theory and aesthetics. The author's intention might well be relevant for interpretation, as that discourse is concerned with the correct or appropriate meaning of a work. However, our question is whether an intention (in the ordinary sense) is a necessary condition for intentionality. I argued that it is not, and characterised instances of intentionality as merely 'potentially meaningful'. The second disadvantage of the concept of derived intentionality has to do with the adjective 'derived' itself. I examined cases of diverging intentionality and differentiated between content derivation and function derivation. The strong claim that the content of a sentence is 'securely derived' from a thought content, or that an audience derives thought contents 'in suitable ways' from artefacts, cannot be maintained across all instances. Finding instances of intentionality in different realms, for example, in mental states and in artefacts, does not imply that we can securely derive the content of one from the corresponding other. We can only sustain the weaker claim of function derivation: that artefacts derive their capacity to represent – that they can fulfil the function of being directed to or 'about' something – from the intentionality of the mind.

Because intentionality is typically only attributed to mental states, we must ask whether it is justified to extend the notion to artefacts.To approach this question, what we mean by intentional properties must be specified. If we intend to attribute derived intentionality to artefacts, we must carefully distinguish it from intensionality, as well as from both meaning and reference. I further differentiated between content derivation and function derivation, and between intentional properties and (intrinsic) intentionality. With these distinctions in place, I proposed to define intentional properties thus:

(ip) An intentional property is a property that indicates contentfulness.

This is not intended to mean the particular content but rather the capacity to have content at all:the capacity to be about something rather than that which something is about. In this sense, it would at least not be incorrect to attribute intentional properties to certain physical items and events.

Finally, I considered three options for question (B)":

ms ----- o
ms ----- a ----- o
ms ----- a (o)

The first option assumes that when we read a book or watch a film, our mental states are directly related to the intentional object. The second one (which I attribute to Searle) acknowledges that the intentional object is mediated through a form of presentation. I am opting for the third possibility, which emphasizes that the intentional content is 'fixed' in a form of presentation. That is, we have direct access not to the intentional content but only to its form of presentation. So, whereas in the second case it seems that we simply must get the meaning right, in the third case the gulf between mental state and represented content is insisted upon. With the distinction between function and content derivation I argue that when we encounter intentional properties in physical items, they indicate that there is content; however, this does not guarantee secure content derivation. If we decide to attribute intentional properties also to physical items or events, then in the description *a(o)*, with *a* standing for the physical item or event and *o* for the intentional object or content, this could denote the parenthesis (...): not the artefact nor the object/content, but rather that which indicates that an item has a content.

In chapter 4, I finally turned to what I have called the 'general' understanding of intentionality, which often remains implicit. I set out to use the paradox of fiction as a tool to analyse the concept of intentionality. What has emerged is a landscape marked by divergent, and potentially incompatible, conceptions of the term. This is perhaps the most significant, and in some respects concerning, outcome of my analysis. These different 'general' understandings of intentionality must be taken seriously. To navigate the vast domain of intentionality, I organised my analysis around commonly used paraphrases of the term. I identified (some) of the paraphrases used to describe intentionality: namely, intentionality as *content* and *directedness*, and as *property of mental states* or as *capacity of the mind*. The analysis of these paraphrases provided the structure of the chapter.

Regarding the description of intentionality as *content and directedness*, I examined various passages from Brentano in which he articulates his notion of *intentional inexistence*. I concluded that Brentano needs both notions for his concept of intentionality. Notably, he does not simply relativise them but rather views each through the lens of the other. On the one hand, the idea that a mental state can 'contain' or 'have' something as its object is explained in terms of its relation to that object. Here, 'object' is used in the grammatical sense: a thought has something *as* its object. For a thought to have an object means that a subject is mentally relating to something, regardless of whether that object exists. On the other hand, a relation is defined such that 'the one who is thinking must certainly exist, but the object of his thinking need not exist at all'. This specific relationship is further characterised by the way we make something an object of thought: namely, by directing ourselves mentally towards it. The mutual explanation of the terms 'object' and 'relation' reveals that Brentano requires both content and directedness to adequately describe the concept of intentionality. I concluded that the simultaneous descriptions of intentionality in terms of content and directedness are not problematic; they do not represent two different understandings of intentionality but rather two complementary aspects of a single phenomenon. This twofold description is what I refer to as the *twofold nature of intentionality*: our mental states have content by being directed towards something, and directedness is explained in terms of 'having an object'.

Concerning the second differentiation – the paraphrasing of intentionality as a 'property of mental states' or as a 'capacity of the mind' – I then returned to the example of Döring, Helm, and Slaby. Although all are proponents of affective intentionality, their general understanding of intentionality diverges. Döring and Helm understand intentionality as a property of mental states, something they share with the proponents of the propositional model of intentionality, which has been influential in emotion theory. However, they do not take intentionality to consist in having a propositional attitude. I have shown that the understanding of intentionality in early cognitive emotion theory, especially in the work of Kenny, was shaped by his engagement with Chisholm's account. This helps explain the tendency to conceive of intentionality as a propositional attitude in that context. For Slaby, on the other hand, intentionality is a mental activity: a 'behaving-towards' something that is experienced as relevant and thus enables us to grasp what is going on around us. This mental activity can, but does not have to, be instantiated in a propositional form. He does not primarily understand intentionality as a property, but as an activity. This marks a fundamentally different understanding of intentionality, and using the same term for both can amount to a case of equivocation if they are not otherwise differentiated. I further considered the possibility of understanding intentionality as the 'capacity of the mind to have intentional states', and the possibility that the distinction between property and activity is already addressed in the difference traditionally drawn between mental states and mental acts. However, neither of these possibilities can account for the divergences found between these different notions of intentionality. The difference between mental states and mental acts may, in fact, actively contribute to further misunderstanding. To differentiate these two accounts more clearly, I decided to refer to the latter as intentionality as an *activity of the mind* – with activity understood as distinct from both capacity and mental acts. Finally, I traced both aspects – of property and activity – in Brentano's writings. I suggested that this distinction is most evident in his so-called dual relation thesis (the claim that 'every mental activity is accompanied by a consciousness which refers to it'): namely, the simultaneous occurrence of the mental activity of *directedness* and the acknowledgement of the very same act *as being directed*. However, Brentano is primarily interested in reconciling these two aspects of intentionality to maintain the unity of consciousness. Moreover, I showed that his complex notion of the *psychische Beziehung* has been flattened through translation. I argued that the notions of intentionality as property and activity must be differentiated, since otherwise this would result in equivocation. A clear distinction can furthermore provide a framework for addressing cases in which these notions have been conflated, as well as elucidating Brentano's complex account and inviting a deeper appreciation of how mental phenomena relate to their objects.

Whereas the twofold description of intentionality as 'content' and 'directedness' is unproblematic and helps to explain its twofold nature, the equivocal use of intentionality as 'property' and 'activity' may lead to category mistakes and conceptual confusion. It is therefore necessary to distinguish these two uses of the term 'intentionality'. However, rather than discarding one or the other, it may be fruitful to inquire in what relation they stand to each other.

5.2 Requirements

In this study, I have examined intentionality through two perspectives offered by emotional responses to fiction. This did not aim to solve the paradox, but rather to analyse the concept(s) of intentionality involved. The insights gained from this analysis can now be formulated as requirements that atheory of intentionality should be able to account for.

(1) A theory of intentionality must account for the intentionality of emotions. Emotions are often experienced and described as being directed towards, or 'about', something. Moreover, the distinction between thoughts and emotions cannot be sharply drawn, and it seems difficult to identify instances of 'pure' thought or 'sheer' emotion in our mental lives. Notably, there is no change in our intentionality whether we have a thought or an emotion about something. We may entertain thoughts about it in one moment and empathize with it in the next. It is not our relation to the object that changes, but rather how we relate to it.
(2) A theory must acknowledge the inseparability of phenomenality and intentionality in emotional experiences. It should demonstrate that emotions cannot be reduced to mere judgments or beliefs but must be understood as unified experiences involving both intentional content and phenomenal quality. The unification view (UV) helps us adequately individuate an emotion, as we need to distinguish both a headache from indignation, and the emotionless judgment that a situation is dangerous from the experience of panic. The related aspect of inferential encapsulation proved particularly useful in explaining the phenomenon of 'conflict without contradiction' in emotions.

(3) A theory of intentionality must distinguish between the correctness conditions for emotions and beliefs/judgements. Due to their truth conditions, beliefs and judgements require an existential commitment to their intentional object. They are thoughts with a mind-to-world direction of fit and therefore necessarily imply an existential commitment. Different correctness conditions apply to emotions. While the formal object of a belief or judgement is truth, the formal objects of emotions differ – the formal object of fear, for example, is danger, and that of indignation is injustice. Thus, beliefs necessarily involve an existential commitment, but emotions do not. This implies either that not all intentional states have a direction of fit, or that direction of fit must be conceived differently in the case of emotions.
(4) If we intend to attribute intentional properties to physical items and events, a theory of intentionality must carefully distinguish these properties from the concepts of meaning and reference. The intentionality of artefacts is only potentially meaningful – it requires a viewer or reader to attribute meaning to them. Similarly, artefacts are only potentially referential – reference occurs only when the object of the mental state exists in reality. We could formalise an artefact with the description *a(o)*, where *a* is the physical item or event and *o* is the intentional object or content. The content can be related in two ways: intensionally and extensionally – that is, *o* may, depending on the context, be understood as the reference or the meaning of the artefact. Neither meaning nor reference should be denoted by the concept of intentional property. Rather, what it can denote is the parenthesis *(…)* – that which indicates that an item has content. How such properties can be individuated is one of the tasks a theory of intentionality must address. To avoid conflating intentional properties with meaning or reference, we can understand them as (ip) 'properties that indicate contentfulness'.
(5) If we attribute intentional properties to physical items and events, a theory of intentionality must distinguish between content derivation and function derivation. The intentional properties of artefacts should be understood as derived from the intentionality of the mind in the weak sense of function derivation, which suffices, although content derivation remains possible. When we read a book or watch a film, we have 'direct' access not to the intentional content but only to its form of presentation: we are confronted with the intentional properties of an artefact, which do not guarantee secure content derivation. In other words, it cannot be assumed ei-

ther that the exact content intended by the author can be fully derived, or that the author succeeded in imposing the intended content on the artefact. Moreover, in some cases, the intentional properties we encounter in physical items and events are underived, yet they can still serve the same *function* as derived intentional properties: we can confer meaning on them or *use* them meaningfully.

(6) If we attribute intentional properties to physical items and events, we must further specify the role of their form of presentation. By 'form of presentation', I refer to the various aspects that pertain to the realisation of an artefact. It corresponds neither to Searle's 'illocutionary force' nor to Frege's 'mode of presentation'; however, it entails aspects of both. We necessarily encounter instances of intentionality in the physical realm through a form of presentation, which involves both logical properties and material aspects.Although physical presentation (words uttered, sentences written, images drawn, etc.) gives us access to content, it does not guarantee secure content derivation. Presentation inevitably contributes to the meaning of an artefact; however, we observe an asymmetry: once intentional content is 'fixed' in a form of presentation, an item becomes potentially meaningful, while the content becomes ambiguous.

(7) A theory of intentionality must account for what I call the 'twofold nature of intentionality': the relationship between directedness and intentional object. Brentano relativised both the terms 'object' and 'relation' and explained each by reference to the other. He needed this mutual explanation to adequately describe intentionality: mental states have content by being directed towards something, and directedness is explained in terms of 'having an object'. A theory must acknowledge both aspects as essential for intentionality; it cannot be reduced to either.

(8) A theory must clearly distinguish between intentionality as a property attributed to mental states and as an activity of the mind. These are instances of fundamentally different kinds, and a clear distinction is necessary to avoid equivocation and category mistakes. Moreover, differentiating between them provides not only a framework for addressing cases in which these notions have been conflated, but also for inquiring into what relation might obtain between them. This, in turn, can foster a deeper appreciation of how mental phenomena relate to their objects.

5.3 Intentionality as *relatedness*

As though there were not already enough ways of thinking about intentionality, I wish to propose yet another. What follows is not a developed theory but rather a possibility for understanding the term that emerged over the course of this analysis. I did not set out to develop a new account; I merely chose not to ignore or dismiss the view that arose. It took shape and gradually clarified in response to the various instances, challenges, and requirements encountered throughout this study. What I propose is merely a new paraphrase of intentionality; nevertheless, I believe it is crucial to remain attentive to its complexity and implications, in order to refine an adequate theory of intentionality.I will briefly present it and indicate some of the points in my analysis with which it engages. I hope that it may prove useful for further consideration.

Relatedness

The understanding of intentionality that emerged during this study is best captured by the term 'relatedness'. The verb 'to relate', derived from the Latin *relatus*, encompasses several meanings: it can mean 'to stand in some relation' or 'to bear reference to something'; alternatively, 'to recount' or 'to tell', as in the phrase 'she relates what happened to them'. This allows relatedness to be used for both mental states and speech acts or other artefacts; moreover, it can denote both an activity and a property – 'she relates a story' and 'the story is related to what happened'. Notably, 'to relate' is also used to express a sense of connection or sympathy. When we say, 'I can relate to that', we express understanding or empathy for another's experience or point of view. This can be affective, but it need not be; it can simply indicate comprehension of a behaviour or situation, while retaining the potential for affect. In short, intentionality concerns *how we relate*. Relatedness can be understood as the nominalisation of the verbs

- to stand in relation, to refer
- to narrate
- to be understanding or sympathetic with.

By encompassing these various meanings, the paraphrase serves as a reminder that each of them must be taken into account in an adequate theory of intentionality, as must their interrelation.

Brentano's *Beziehung*

One consideration that may justify the use of the term 'relatedness' is Brentano's description of the intentional relation as *psychische Beziehung*. I have remarked on this term at various points throughout this study, particularly with regard to how its translation as 'mental reference' flattens Brentano's original notion and obscures important aspects of his understanding of intentionality. Translating *Beziehung* as 'reference' is misleading, since Brentano did not intend to imply the strong sense of *direct reference*, where reference fails if the referent does not exist. As Taieb (2017) points out, Brentano clearly distinguished between intentionality and reference – the latter obtaining only when the object of a mental state exists in reality. For Brentano, the intentional relation can, but does not necessarily, involve an existential commitment. By 'Beziehung auf einen Inhalt', he means only that something is presented in thought – in the sense of being 'put before the mind'. Furthermore, as I have argued in 4.3.4, Brentano describes intentionality essentially as an activity. 'Alles psychisch sich Beziehende' (1925, 158), for instance, literally translates as 'everything mentally relating', but is standardly translated as 'all mental references' (1995, 227). Translating *Beziehung* as 'reference' not only privileges a linguistic reading of intentionality but obscures the aspect of mental activity that recurs throughout Brentano's writings. It thus suggests a reductive sense of intentionality.

Similarly, translating *Beziehung* as 'relation' also introduces difficulties. Brentano distinguished the intentional relation from the general philosophical notion of relation, which in its classical sense typically presupposes the existence of both relata.[1] By contrast, in an intentional relation, only the fundament (the one who is thinking), not the terminus (the object of thought), must exist. Brentano dedicates an entire chapter to clarifying this distinction in the appendix to the 1911 edition. Here, he famously describes the intentional relation as something 'quasi-relational' (1995, 212). His understanding of intentionality as a special kind of relation is complex and Taieb identifies two major shifts in Brentano's thinking: first, a move from his early conception of immanent objects to a quasi-relational account of intentionality, motivated

1 This understanding goes back to the Aristotelian tradition, in which relations are conceived as holding between actually existing terms, whereas some approaches in modern logic allow for quantification over nonexistent or fictional objects (cf. Linsky 1967; Lambert 2002) or, indirectly, in possible-world semantics (e.g., Kripke 1980).

by the realisation that the former would isolate us from access to the external world; and second, a return to a relational understanding of intentionality – one that allows for the existence of relations even when both relata do not exist – while still rejecting the existence of immanent objects.[2]

This complexity is reflected in Brentano's use of the German term *Beziehung*. It can be translated as either 'relation' or 'reference', though both terms tend to imply an existential commitment, which Brentano did not intend. While there is a tendency towards 'reference', as *Beziehung* is often used in the sense of *sich beziehen*, his emphasis is placed on the mental activity of referring rather than on the object being referred to. This sense would nowadays be better captured by the term *Bezogensein*, and verbs such as 'attending to' or 'relating' more accurately convey the mental activity Brentano had in mind. Given the connotations of 'relating' – including the senses of 'referring', 'narrating', and 'being sympathetic or understanding' – the term 'relatedness' seems the most apt way of paraphrasing Brentano's notion of intentionality. Moreover, as a nominalisation of the verb 'to relate' that expresses the state or condition of relating/being related, it avoids some of the more technical implications typically associated with the term 'relation'.

Thoughts and emotions

One advantage of conceiving intentionality as relatedness is that it obviates the need to posit a distinct kind of intentionality for emotions. Furthermore, this conception allows us to incorporate the 'unification view', which holds that phenomenality and intentionality are inseparable in emotional experience. This addresses requirements (1) and (2). Importantly, the term suggests that affect is not merely added onto a belief or desire; rather, relatedness *in itself* bears the potentiality for affect. We do not relate to everything we perceive, nor does everything we relate to possess phenomenal qualities. Moreover, not everything we relate to prompts emotional responses; we may, for example, hold nonaffective mental states like beliefs grounded in the way we relate to something. I have argued that while we must account for the intentionality of emotions, it is preferable not to posit a different kind of intentionality for this purpose. There is no change in the intentional relation according to

2 Cf. Taieb (2017). For an in-depth study of Brentano's notion of relation and its development across his philosophical work, see Taieb's book *Relational Intentionality: Brentano and the Aristotelian Tradition* (2018).

whether we have a thought or an emotion about, say, Frankenstein's monster. We might entertain thoughts about the possibility of Frankenstein's monster in one moment, empathise with it in the next, and at the same time reflect on the formal aspects of how Mary Shelley evokes our empathy for the monster. In none of these cases does our relation change – our mind remains directed towards Frankenstein's monster – only how we relate to it. The *mode* of relating may differ depending on whether we believe, desire, or feel something towards it, but the *relatedness itself* remains unchanged.

At this point we can recall Brentano's dual relation thesis, particularly his emphasis on the temporal congruity of being directed towards an object and what he calls 'inner consciousness', that is, the acknowledgment of that very same activity. He insists that these are not separate, but rather parts of the same mental phenomenon. This point of Brentano's could provide a strong basis for the unification view, as perception and its acknowledgment (or its noninferential judgement) are conceived as aspects of the same unified mental phenomenon. In terms of affective intentionality, being frightened, for instance, is the assessment of a situation as dangerous through the phenomenal experience of fear. The dual relation thesis allows us to explain how this assessment is not 'added on', but immediate and noninferential. In this sense, phenomenality and intentionality would be united in the conscious experience. Moreover, this avoids what Helm calls the *problem of emotionality*: where should the specifically emotional aspect of an experience come from if its core, the intentional relation with something in the world, is not already (potentially) emotional or affective? Thus, the dual relation thesis also helps identify emotions non-circularly.

However, Brentano's view is not limited to emotions: according to him, every mental state bears this dual relation.[3] Textor's (2013) reading of Brentano is particularly interesting because he proposes understanding the relation between the activity of directedness and its acknowledgement as one in which the latter *individuates* the former. This interpretation may help satisfy requirements (1) and (2) without the need to posit a different kind of intentionality for emotional responses. If we understand intentionality as a mental activity that is durational and distinct from mental acts (as pointed out in 4.3.3), it can

3 One should bear in mind that Brentano allows for unconscious mental states in the form of dispositions, but does not include them in his analysis (cf. chapter 4.3.4). Such cases might, arguably, be conceived as acknowledgements that are not actively entertained, and thus would not fall under the dual relation thesis.

be argued that we respond to shifting perceptual stimuli and new information over time, and consequently individuate distinct mental states, while what we are relating to remains constant. We may feel fear of Frankenstein's monster in one moment and curiosity or empathy for it in the next. As long as our attention remains directed towards the same object, our relatedness does not change – only how we relate changes.

Divergent correctness conditions

Nevertheless, we must account for the fact that the correctness conditions of emotions differ from those of beliefs or desires: their formal object is not truth, but, for example, danger or injustice (cf. Döring 2009a). While propositional attitudes, such as beliefs, are assessed in terms of truth (a belief is correct if the proposition it affirms is true), emotions are assessed in terms of appropriateness. Fear is appropriate if something appears to pose a threat. While the intentional object in a proposition is thus related to the existence of the object or state of affairs referred to, the intentional object of an emotion (or other mental state without direction of fit) depends rather on how something appears to us, or under what aspect it is presented. Accounting for these different correctness conditions, formal objects, and directions of fit forms requirement (3).

In this regard, it is worth recalling that for Brentano the most 'basic' intentional mental states are presentations, and only when they are accompanied by an evaluation do they become, for instance, a belief or judgement. In his view, 'presentations are the foundations of all other mental phenomena. [However,] mental phenomena are [not] merely objects of presentation [,but] they can be in our consciousness in still other ways. What is certain is that knowledge frequently accompanies mental phenomena. We think, we desire something and know that we think and desire. But we only have knowledge when we make judgements. It is beyond doubt, therefore, that in many cases *along* with the mental act there exists within us not only a presentation which refers to it but also a judgement about it.' (Brentano 1995, 107, my emphasis). While beliefs and judgements may involve an existential commitment, presentations do not; presentations are merely acknowledged (*anerkannt*) by the mind, not affirmed or denied. To entertain a thought about Frankenstein's monster – imagining for instance what it looks like – is distinct from believing (or denying) that it existed, and has no direction of fit. In this sense presentations are for Brentano the foundation of all other mental phenomena. We can observe a certain analogy to Searle's categorisation of illocutionary acts: he claims that next to assertives, directives, commissives, and declarations, there are also expressives, which, unlike the others, do not have a direction of fit.[4] Furthermore, only speech acts and mental states with direction of fit, are truth-evaluable and may involve existential commitment.

The notion of relatedness can help us here, since it preserves the idea that the intentional object is constituted by being something both related to (that is, experienced in a particular way) and related (that is, presented under some aspect). The former can be described as a non-inferential acknowledgement of the intentional object, which does not yet imply an existential commitment to it, nor an attitude towards the truth value of a proposition. However, it involves the object's appearing under a particular aspect. After judgements, Brentano continues to specify the conditions of the phenomena of love and hate. Similarly, these are based on presentations: one cannot love, hate, fear, or admire something unless that thing is first presented to consciousness. In this sense, presentations are the most basic intentional mental states, while emotions are more complex ones founded upon them. Just as judgements are subject to standards of truth and falsity, phenomena of love and hate are subject to standards of correctness. According to Brentano, emotions are not merely subjective feelings but intentional acts that evaluate their objects. An emotion is correct when it is directed towards an object that merits the corresponding positive or negative response. In this way, truth is the norm governing judgement, while value is the norm governing emotions. Brentano further argues that values can be apprehended with a form of evidence analogous to that found in self-evident judgements. Consequently, emotions do not simply express preferences; they can disclose objective value and provide the basis for ethical knowledge.[5] A detailed comparison of his account of emotions with contemporary debates on affective intentionality lies beyond the scope of the present study, although such an investigation would undoubtedly prove fruitful. Importantly, he acknowledged the different correctness conditions of beliefs and emotions, while grounding both in the more fundamental intentional phenomenon of presentation.

Brentano's conception of presentation (*Vorstellung*) as the fundamental mode of intentional mental states, on which judgements and emotions are

---

4 Cf. Searle (1983, 166), and in the essay 'Expression and Meaning' (1979a).

5 Cf. the second volume of Brentano's *Psychology*, particularly chapter 7 and 8. There, he develops further distinctions between beliefs and judgments, as well as between emotions and desires, as part of a comprehensive theory of mental categorisation.

founded,[6] already accounts for the condition of possibility of there being different correctness conditions and formal objects for emotions and other mental states without direction of fit.

Content and directedness

Furthermore, the notion of relatedness retains and makes legible what I have termed the *twofold nature of intentionality*. Brentano requires both *content* and *directedness* to explain his concept of intentionality, particularly the notion of intentional inexistence. He not only relativises the notions of 'object' and 'relation' but also uses them to mutually explain one another (as discussed in 4.2). Relatedness holds both aspects: that something becomes a (grammatical) object by being related to (in the sense of being referred, narrated, or having import), and that if we relate, we necessarily relate *to* something. The notion of directedness does not, for instance, imply the latter: we can be directed without being directed toward something. Accounting for this twofold nature of intentionality would address both requirements (7) and (3).

The descriptions of intentionality in terms of content and directedness provide a framework for situating some of the understandings of intentionality encountered in this study. Both propositional intentionality and what I have called 'objectual intentionality' emphasise the intentional object. They identify an intentional state by it having an intentional object and specify different ways in which we relate to it. Having a propositional attitude corresponds to what Searle terms the 'intentional mode', claiming that every intentional state is composed of a propositional content and an intentional mode, thereby allowing, for example, a distinction between the belief 'that it rains' and the hope 'that it rains'. Here, the direction of fit determines the truth conditions of the proposition. However, this schema does not apply to all intentional states. In emotions, for instance, our relatedness is not necessarily expressed as an attitude towards a proposition, but rather by means of the presentation of the intentional object in a certain way – as frightening, admirable, or lovable, etc. This is why proponents of affective intentionality argue against a strictly propositional understanding. Many emotions relate us to objects or situations in ways that cannot be reduced to propositional content. Here, the intentional

6 Although this may suggest an 'add-on' view similar to early cognitive theories of emotion, Brentano emphasises the unity of consciousness in these different phenomena. Cf. chapter 4 in the first volume of his *Psychology*.

object is presented under a particular aspect, but not necessarily in the form of a proposition. When we are afraid at the sight of a snake, we do not necessarily entertain the proposition 'that that snake is dangerous'; rather, we experience it *as dangerous*.

Consequently, many proponents of affective intentionality do not understand intentionality in propositional terms. Rather, many conceive it as what I have called 'objectual intentionality', identifying intentional states by their 'having an intentional object'. Others, however, go further and argue that 'having an object' is not a necessary condition for affective intentionality and instead emphasise the aspect of 'directedness' (cf. Slaby). The notion of relatedness can help avoid certain problems that come up in this debate. 'Being directed' is, in itself, too vague a concept. For Brentano, directedness is shorthand for being 'directed toward an object'. His mutual explanation of the terms 'object' and 'relation' reveals that he requires both content and directedness to adequately describe intentionality. This is the *twofold nature of intentionality* (as discussed in 4.2.3): our mental states have content by virtue of being directed towards something, and directedness is explained in terms of 'having an object'. We do not relate to everything we perceive; something must bear some import for us in order for us to direct our attention to it. Content and directedness depend on each other in the intentional relation: by directing our attention towards something, we individuate a content, and our attention is drawn to something because it is of importance to us. In this sense, content and directedness mutually constitute each other.

The notion of relatedness captures this specific connection. While we are, by virtue of our bodily situatedness in the world, always directed towards things, we relate only to some of them. Therefore, conceiving of intentionality solely as directedness, devoid of an intentional object, fails to do justice to the specificity of the intentional relation. However, there are cases where individuating the object we relate to may be difficult or require time and additional information. For example – and to return briefly to emotional responses to fiction – it is a common technique in horror movies to obscure, or not clearly show, the object of fear. Feelings of suspense or even fear arise despite, and even because of, this ambiguity. These feelings are directed towards something, albeit something unknown, vague, or only guessed at. This prompts us to imagine all kinds of objects, yet we cannot be certain which of these horrific things might await around the corner. This example illustrates that, although a mental state may lack a clearly defined intentional object, the intentional relation nevertheless calls for one: it implies a tendency to identify an object to which we relate.

We cannot conceive of intentionality merely as directedness: if we emphasise this aspect, it must at least involve the tendency to obtain an object. Similarly, intentionality cannot be reduced solely to content. Something is only an intentional object insofar as it is specified how we relate to it – that is, by its intentional mode or the manner in which it is presented. Neither 'content' nor 'directedness' alone can fully capture the intentional relation. Using the term 'relatedness' is advantageous in this respect: it suggests the activity of 'being directed', yet unlike the notion of direction, which can in principle be conceived of without its having an object, 'relating to' indicates a more determinate connection. Relatedness implies that we relate to something.

Artefacts

As I mentioned when introducing the term *relatedness*, the verb 'to relate' can have several meanings. The two primary meanings in English are (1) to 'stand in a relation' or 'be connected' (e.g., 'These ideas relate to each other'), and (2) to 'narrate' or 'report' (e.g., 'she related the story in detail'). While these uses may seem distinct, they share a common etymological root in the Latin verb *referre* (from *re-* 'back' and *ferre* 'to carry'), which means 'to bring back', 'to report', or 'to refer'. The past participle *relatus* gave rise to the English *relate*. Both senses preserve the core idea of bringing something into connection. In the first, one item or idea is connected to another; in the second, an event or experience is brought into discourse and shared. The concept of *relatedness* draws on both these dimensions. This allows relatedness to be used in addressing both the intentional properties of mental states and those of speech acts or other artefacts.

However, this may be no more than a historical connection, contingent on how the term developed in different contexts, and need not imply a genuine conceptual link between the intentional properties of mental states and those of artefacts. Clearly, it does not resolve the question of whether it is useful to attribute such properties to the latter. At the end of chapter 3, I defined intentional properties in such a way that it would be at least not incorrect to attribute them also to physical items and events. We must, however, be careful not to conflate intentionality with intensionality. Furthermore, I have distinguished intentionality from both meaning and reference, and differentiated between content derivation and function derivation, between derived and underived properties, and between intentional properties and (intrinsic) intentionality. These aspects are outlined in requirements (4) and (5). With these distinctions in place, we can understand intentional properties as *properties that indicate contentfulness*.

I want to suggest two potential advantages for attributing intentional properties also to physical items. Firstly, it may provide a more suitable and expansive vocabulary for analysing AI-generated content. As AI systems increasingly produce outputs that resemble human-generated language, images, or decisions, there arises a need to account for their apparent 'aboutness' without presupposing mental activity or intrinsic intentionality. While these systems lack mental states, their outputs are routinely interpreted as meaningful or goal-directed within specific contexts. Moreover, in allowing for instances of unintentional and diverging intentionality (as argued in 3.3.3), this pragmatic treatment suggests that intentional attributions may be functionally grounded rather than ontologically committed. Treating the intentionality of artefacts as a derived property – grounded in function and use, rather than origin or reference – may help us describe the role such artefacts play in human cognitive and communicative practices. This avoids the category mistake of attributing mental states to nonconscious systems while preserving a framework for analysing how such artefacts function within communicative practices in which meaning is distributed among human and nonhuman agents. Moreover, AI-generated content may help us individuate intentional properties.

Activity and property

Secondly, the concept of derived intentionality might help us think about the difference between understanding intentionality as *activity* and as *property*. The derived intentional properties of physical items could be considered a 'plain' instance of an intentional property. For example, the sentence 'It is raining' literally means that it rains, but it only actually refers to something if it is instantiated by someone – either as a mental state (such as believing, hoping, or fearing that it rains) or through a physical performance (such as stating that it does). As a property of a sentence, it is disconnected from the mental activity of relating: it is neither related to something (in the sense of referring) nor itself related (in the sense of narrated or otherwise enacted). In this respect, it constitutes a sheer property. And it would be absurd to attribute mental states to the words written in a textbook. Searle's distinction between derived and intrinsic intentionality is modelled primarily on a division between different realms, and he does not explicitly address the distinction between intentionality as property or activity.[7] That distinction, which I propose, instead shifts the focus from different realms to different types of mental phenomena. In this sense, mental states and physical items exhibiting intentional properties are closer to one another than either is to the mental activity of intentionality. Of course, important differences remain between mental states and artefacts – particularly regarding meaning, intention (in the ordinary sense), and their production or use. Requirement (6), for instance, asks us to specify the functional role played by the *form of presentation*, and to account for the peculiar asymmetry that arises when a content is 'fixed' within a particular form of presentation. Nevertheless, the structural resemblance between the intentional properties of mental states and artefacts is greater than that between either and the mental activity of intentionality: the former are representations, whereas the latter is not.[8]

If this line of thought is further pursued, intentionality as activity would only apply to occurrent mental states, whereas intentional properties could be attributed to both (occurrent and dispositional) mental states and artefacts. Although this distinction is different from Searle's, conceiving of derived intentionality in artefacts as sheer properties – clearly distinct from and devoid of intrinsic intentionality – may help clarify the relationship between intentionality as a *mental activity* and intentionality as a *property of mental states*. Insisting on this distinction is, I believe, one of the central outcomes of this study. When the term 'intentionality' is used equivocally, it risks generating category mistakes and conceptual confusion. A thorough examination of this distinction, as well as of the relationship between these forms of intentionality, is still needed. Here, I can only offer some preliminary suggestions.

How might the difference between intentionality as property and activity be conceived within the mental realm? Beyond demanding a clear distinction between property and activity, requirement (8) also calls for an account of the relationship between these two kinds of intentionality. I traced some aspects of these different notions in Brentano's work (cf. chapter 4.3.4); his dual relation

7 The relationship between what Searle calls 'primitive' intentionality and representational states – whether mental or illocutionary – might nevertheless be worth exploring.

8 This, furthermore, relates to the notion of direction of fit: only representational states have a direction of fit because the 'fit' requires a distance between the content of the state and its object.

thesis, in particular, offers a possible point of departure for conceptualising this distinction. In this thesis, we could take 'directedness toward an object' as the *activity* of intentionality, and what he calls 'inner consciousness', that is, the acknowledgment of this directedness, as the intentional *property*. Concerning the former, we can recall Brentano's emphasis on the notion of 'mental activity' – what he called *psychische Beziehung*, and could perhaps be better expressed today by the term *Bezogenheit*. The latter notion is more complex and needs interpretation. It is disputable how this mental phenomenon that happens *en parergo* – this immediate and noninferential acknowledgement of the mental activity – can be explained. If we follow Textor's (2013) reading of Brentano, the relation between the activity of directedness and the acknowledgment of that very activity is such that the latter *individuates* the former. In this sense, the two may be regarded as distinct mental phenomena: one of directedness, the other of individuation. The 'result' of individuation – the mental state individuating the object of directedness – can thus be attributed the property of being directed. Up to this point, the account is congruent with Textor's reading. We may, however, take it a step further. The concept of derivation may help to further clarify this dual relation: intentional properties could be understood as derived from the mental activity of intentionality. In this case, however, derivation would have to be conceived not as an act (as in Searle's account), but as something unintentional (in the ordinary sense) and automatic. The relation between property and activity may thus be understood as *automatic derivation*, in which the latter unintentionally, but concomitantly, individuates the former. This would account for the fact that the resulting mental state represents the corresponding mental activity. Furthermore, whereas in artefacts only function derivation securely applies, the dual relation of mental states also allows for secure *content derivation*. Given the unity of consciousness that Brentano insists upon in his dual relation thesis, a temporal congruity between activity and property is assumed within the mental realm. This applies to all occurrent mental states: whether we form a new thought or recall a dispositional content, we are securely 'connected' (or re-connected) to the object of that thought while entertaining it. By contrast, the intentional content of an artefact is disconnected from its intentional object and therefore always requires interpretation. This difference provides a possible starting point for addressing requirement (6). In this sense, intentional properties may be conceived as having been derived from the mental activity they represent: individuation attributes an intentional property to a mental state by specifying what the state is 'about'. Thus, the notion of derived intentionality may offer a useful model for understanding the relationship between intentionality as activity and as property.

Nowadays, the understanding of intentionality as a property of mental states is prevalent and both propositional intentionality and what I have called objectual intentionality fall under this conception. However, reducing intentionality to being solely either a property or an activity is insufficient. Brentano emphasised that the mental activity of directedness and its acknowledgment are inseparable in what he calls the unity of consciousness. His point was well taken since tying intentional content to the activity of directedness avoids treating the object as an independent entity. Nevertheless, as this study has shown, it remains crucial to distinguish intentionality as activity from intentionality as property in order to avoid equivocation and category mistakes. By understanding intentionality as *relatedness*, I suggest we can preserve the unity of the phenomenon while analytically distinguishing the aspects of activity and property and exploring their interrelation.

5.4 Conclusion

Examining intentionality through the two perspectives offered by emotional responses to fiction proved useful to further refine the concept. It revealed implications and inconsistencies in the use of the term and, in particular, enabled the formulation of requirements (1)-(8) that an adequate theory of intentionality should satisfy. Notably, we must be careful not to conflate intentionality understood as a property of mental states with intentionality understood as an activity of the mind. While the analysis was restricted to specific examples from the debates around the paradox and to the concepts of affective and derived intentionality, the intersection of these perspectives has revealed not only inconsistencies, but also potential ways of conceiving intentionality more fully and adequately. The paraphrase *relatedness* can further help us remain attentive to the complexity and implications of this concept, as it draws on the various senses of the verb 'to relate' – including referring, narrating, and being sympathetic to something. Furthermore, it denotes the state or condition of relating and thus avoids the implication that intentionality must necessarily be conceived as a dyadic relation. By encompassing these various meanings, *relatedness* serves as a reminder that each of these aspects must be taken into account: intentionality concerns *how we relate*.

Epilogue

> So writing is the method of using the word as bait: the word fishing for whatever is not word. When this non-word – between the lines – takes the bait, something has been written. Once whatever is between the lines is caught, the word can be tossed away in relief. But that's where the analogy ends: the non-word, taking the bait, incorporates it.
> *Clarice Lispector, Água Viva*

The first version of *Água Viva* was completed in 1971, but Lispector reworked the manuscript and published it only two years later. The text, written in the style of a monologue of the author addressing, as if in a letter, someone from her past, is formless and plotless, and appears to consist of her spontaneously writing down anything and everything that comes into her mind. In the introduction by Benjamin Moser, we learn that the seemingly unfiltered brainstorm is not just painstakingly composed, but that Lispector also had more general doubts: she was hesitant because the book did not seem to be 'about' something, as if missing a backbone. This also seems to be reflected in the title: *água viva* literally translates as 'living water' and can be associated with the stream of thought in which the book is written; for a Portuguese reader, however, it primarily refers to a jellyfish – a fish without a spine. What, then, is this invertebrate floating about? Although the first version of the text was entitled *Beyond Thought: Monologue with Life*, Lispector does not in fact seem to be interested in what lies 'beyond'. At several points, she justifies what she has just written through her method of writing ('I don't like what I just wrote – but I'm duty-bound to accept the whole section because it happened to me,' 46). It seems that what interests her is not what the text is about, but the 'about' itself. As if she were placing herself in a position from which to observe how words come to be about something, persisting in the moment when the 'non-word' takes the bait, becoming word. As if she were using writing as a method for spying on how the 'about' comes about.

Bibliography

Anscombe, G. Elizabeth M. (1957). *Intention*. Ithaca, N.Y.: Cornell University Press.

Binder, Thomas (2008). *Franz Brentano und sein philosophischer Nachlass*. Berlin/Boston: DeGruyter.

Beckett, Samuel (1955/2009). *Three Novels: Molloy, Malone Dies, the Unnamable*. New York: Grove Press.

Brentano, Franz (1924). *Psychologie vom empirischen Standpunkt*. Erster Band. Text der 1874 Ausgabe und Nachtrag von 1911. Hamburg: Meiner.

Brentano, Franz (1925). *Psychologie vom empirischen Standpunkt*. Zweiter Band. Von der Klassifikation der psychischen Phänomene. Text der 1874 Ausgabe und neue Abhandlungen aus dem Nachlaß. Hamburg: Meiner.

Brentano, Franz (1928). *Psychologie vom empirischen Standpunkt*. Dritter Band. Vom sinnlichen und noetischen Bewusstsein. Text der 1874 Ausgabe. Hamburg: Meiner.

Brentano, Franz (1933). *Kategorienlehre*. Edited by A. Kastell. Leipzig: Felix Meiner Verlag.

Brentano, Franz (1952–1966). *Die Abkehr vom Nicht-Realen*. Edited by F. Mayer-Hillebrand. Hamburg: Felix Meiner.

Brentano, Franz (1995). *Psychology from an Empirical Standpoint*. Translation of the 1924 edition. Edited by Oskar Kraus. English edition edited by Lynda McAlister. London: Routledge.

Brentano, Franz (2015). *Psychology from an Empirical Standpoint*. Translation of the 1924 edition with an introduction by Tim Crane. New York: Routledge.

Brito, Evandro (2023). 'Franz Brentano and the Philosophical Crisis in the Four Phases of Its History.' *ARGUMENTOS – Revista de Filosofia* 15 (29): 86–97.

Butler, Octavia (2021). 'Octavia Butler: Visionary Fiction.' Podcast with source material. https://www.npr.org/2021/08/25/1030976863/octavia-butler-visionary-fiction-2021.

Carroll, Noel (1990). *The Philosophy of Horror: Or, Paradoxes of the Heart*. New York: Routledge.

Carroll, Noel (1997). 'Art, Narrative and Emotion.' In M. Hjort and S. Laver (eds.), *Emotion and the Arts*. Oxford: Oxford University Press, 190–211.

Carson, Anne (1998). *Autobiography of Red*. New York: Vintage Books.

Cassian, Nina (1992). *Cheerleader for a Funeral*, translated by Brenda Walker. London/ Boston: Forest Books.

Choi, Jinhee (2003). 'Fits and Startles: Cognitivism Revisited.' *Journal of Aesthetics and Art Criticism* 61 (2): 149–157.

Chisholm, Roderick M. (1955). 'Sentences about Believing.' *Proceedings of the Aristotelian Society* 56: 125–148.

Chisholm, Roderick M. (1957). *Perceiving: A Philosophical Study.* Ithaka: Cornell University Press.

Coleridge (1817/1971). *Biographia Literaria: Or, Biographical Sketches of My Literary Life and Opinions.* London: Scholar Press.

Crane, Tim (1992). 'The Non-conceptual Content of Experience.' In *The Contents of Experience.* New York: Cambridge University Press, 136–157.

Crane, Tim (1998). 'Intentionality as the Mark of the Mental.' In A. O'Hear (ed.), *Contemporary Issues in the Philosophy of Mind.* Cambridge: Cambridge University Press, 229–252.

Crane, Tim (2001a). *Elements of Mind: An Introduction to the Philosophy of Mind.* Oxford: Oxford University Press.

Crane, Tim (2001b). 'Intentional Objects.' *Ratio* 14 (4): 298–317.

Crane, Tim (2002). 'The Intentional Structure of Consciousness.' In A. Jokic and Q. Smith (eds.) *Consciousness: New Philosophical Perspectives.* New York: Oxford University Press, 33–56.

Crane, Tim (2017). 'Brentano on Intentionality.' In U. Kriegel (ed.), *Routledge Handbook of Franz Brentano and the Brentano School.* London: Routledge, 41–48.

Currie, Gregory (1991). 'Imagination and Make-Believe.' In B. Gaut and D. Lopes (eds.), *The Routledge Companion to Aesthetics.* London: Routledge, 335–346.

Davidson, Donald (1970). 'Mental Events.' In L. Foster and J. W. Swanson (eds.), *Experience and Theory*. London: Humanities Press, 79–101.

Davies, Stephen (2009). 'Responding Emotionally to Fictions.' *The Journal of Aesthetics and Art Criticism* 67 (3): 269–284.

De Sousa, Ronald (1987). *The Rationality of Emotion*. Cambridge: MIT Press.

Döring, Sabine A. (2003). 'Explaining Action by Emotion.' *Philosophical Quarterly* 53 (211): 214–230.

Döring, Sabine A. (2007). 'Seeing What to Do: Affective Perception and Rational Motivation.' *Dialectica* 61 (3): 363–394.

Döring, Sabine A. (2009a). 'The Logic of Emotional Experience: Noninferentiality and the Problem of Conflict Without Contradiction.' *Emotion Review* 1 (3): 240–247.

Döring, Sabine A. (ed.) (2009b). *Philosophie der Gefühle*. Suhrkamp.

Döring, Sabine A. (2010). 'Why Be Emotional?' In: *The Oxford Handbook of Philosophy of Emotions.* New York: Oxford University Press, 283–301.

Döring, Sabine A and Müller, Jean M. (2022). 'Emotion as Feeling Towards Value: A Theory of Emotional Experience.' *Mind* 133 (531): 842–851.

Döring, Sabine A., and Reisenzein, Rainer (2009). 'Ten Perspectives on Emotional Experience: Introduction to the Special Issue.' *Emotion Review* 1 (3): 195–205.

Düringer, Eva-Maria (2018). 'Desires and Fiction.' *Journal of Literary Theory* 12 (2): 241–259.

Elgin, Catherine Z. (1999). *Considered Judgment*. Princeton: Princeton University Press.

Frege, Gottlob (1892). 'Über Sinn und Bedeutung.' *Zeitschrift für Philosophie und Philosophische Kritik* 100 (1): 25–50.

Frijda, Nico H., and Scherer, Klaus (2009). 'Emotion Definitions (Psychological Perspectives).' In Sander D. and Scherer K. (eds.), *The Oxford Companion to Emotion and the Affective Sciences.* New York: Oxford University Press, 142–144.

Gallagher, Shaun (2005). *How the Body Shapes the Mind.* Oxford: Oxford University Press.

Goldie, Peter (2000). *The Emotions: A Philosophical Exploration.* Oxford: Oxford University Press.

Goldie, Peter (2002). 'Emotions, Feelings and Intentionality.' *Phenomenology and the Cognitive Sciences* 1: 235–254.

Goldie, Peter (2003). 'Narrative, Emotion, and Perspective.' In Matthew Kieran and Dominic Lopes (eds.), *Imagination, Philosophy and the Arts.* New York: Routledge, 54–68.

Goldie, Peter (2005). 'Imagination and the Distorting Power of Emotion.' *Journal of Consciousness Studies* 12 (8–10): 127–139.

Goldie, Peter (2009). 'Getting Feelings into Emotional Experience in the Right Way.' In S. Döring and R. Reisenzein (eds.), *Perspectives on Emotional Experience. Special Issue to Emotion Review* 1 (3): 232–239.

Goldie, Peter (2011). 'Emotionen, Gefühle und Intentionalität.' In J. Slaby, A. Stephan, H. Walter and S. Walter (eds.), *Affektive Intentionalität: Beiträge zur welterschließenden Funktion der menschlichen Gefühle.* Translation of the revised article from 2002. Paderborn: Mentis, 49–71.

Greenspan, Patricia (1981). 'Emotions as Evaluations.' *Pacific Philosophical Quarterly* 62: 158–169.

Grice, H. Paul (1957). 'Meaning.' *The Philosophical Review* 66 (3): 377–388.

Grice, H. Paul (1969). 'Utterer's Meaning and Intentions.' *The Philosophical Review* 78 (2): 147–77.

Gyekye, Kwame (1971). 'The Terms "Prima Intentio" and "Secunda Intentio" in Arabic Logic.' *Speculum* 46 (1): 32–38.

Helm, Bennett W. (2000). 'Emotional Reason: How to Deliberate about Value.' *American Philosophical Quarterly* 37 (1): 1–22.

Helm, Bennett W. (2001). *Emotional Reason: Deliberation, Motivation, and the Nature of Value.* Cambridge: Cambridge University Press.

Helm, Bennett W. (2002). 'Felt Evaluations: A Theory of Pleasure and Pain.' *American Philosophical Quarterly* 39 (1): 13–30.

Helm, Bennett W. (2009). 'Emotions as Evaluative Feelings.' *Emotion Review* 1 (3): 248–255.

Helm, Bennett W. (2011). 'Affektive Intentionalität: Holistisch und Vielschichtig.' In J. Slaby, A. Stephan, H. Walter and S. Walter (eds.), *Affektive Intentionalität: Beiträge zur welterschließenden Funktion der menschlichen Gefühle.* Paderborn: Mentis, 72–99.

Helm, Bennett W. (2020). 'Affective Intentionality and the Reactive Attitudes.' In T. Szanto and H. Landweer (eds.), *Handbook of Phenomenology of Emotions. Routledge*, 227–238.

Hume, David (1739/40). *A Treatise on Human Nature.* London: Clarendon Press.

James, William (1884). 'What is an Emotion?' *Mind* 9 (34): 188–205.

Jastrow, Joseph (1901). 'Fact and Fable in Psychology.' *The Monist* 11: 290–295.

Kaufman, C., et al. (2018). 'Phenomenal Insight: Pareidolia—I See?' *Australian and New Zealand Journal of Psychiatry* 53 (1): 89–90.

Kavan, Anna (1970/2009). *Julia and the Bazooka*. London: Peter Owen.

Kenny, Anthony (1963). *Action, Emotion and Will.* New York: Humanities Press.

Knapp, Steven, and Michaels, Walter Benn (1982). 'Against Theory.' *Critical Inquiry* 8 (4): 723–742.

Knudsen, Christian (1982). 'Intentions and Impositions.' In N. Kretzmann, A. Kenny and J. Pinborg (eds.), *Cambridge History of Later Medieval Philosophy.* Cambridge: Cambridge University Press, 479–95.

Konrad, Eva-Maria and Petraschka, Thomas and Werner, Christiana (2018). 'The Paradox of Fiction – A Brief Introduction into Recent Developments, Open Questions, and Current Areas of Research, including a Comprehensive Bibliography from 1975 to 2018.' *Journal of Literary Theory* 12 (2): 193–203.

Kriegel, Uriah (ed.) (2017). *The Routledge Handbook of Franz Brentano and the Brentano School*. New York: Routledge.

Kripke, Saul A. (1980). *Naming and Necessity: Lectures Given to the Princeton University Philosophy Colloquium*. Cambridge: Harvard University Press.

Lamarque, Peter (1981). 'How Can We Fear and Pity Fictions?' *British Journal of Aesthetics* 21: 291–304.

Lamarque, Peter (1996). *The Fictional Point of View*. Ithaca: Cornell University Press.

Lamarque, Peter (2005). 'Fiction.' In Levinson, Jerrold (ed.), *The Oxford Handbook of Aesthetics*. Oxford: Oxford University Press, 377–391.

Lamarque, Peter, and Olsen Stein H. (1994). *Truth, Fiction, and Literature: A Philosophical Perspective*. Oxford: Oxford University Press.

Lambert, Karel (2002). *Free Logic: Selected Essays*. Cambridge: Cambridge University Press.

Levinson, Jerrold (1997). 'Emotion in Response to Art.' In M. Hjort and S. Laver (eds.), *Emotion and the Arts*. Oxford: Oxford University Press, 20–34.

Linsky, Leonard (1967). *Referring*. New York: Humanities Press.

Lispector, Clarice (1973/2012), *Água Viva*. Translated by Stefan Tobler. New York: New Directions.

Lyons, William E. (1980). *Emotion*. Cambridge; New York: Cambridge University Press.

Matravers, Derek (2014). *Fiction and Narration*, Oxford: Oxford University Press.

McDowell, John (1994). *Mind and World*. Harvard University Press.

Meinong, Alexius (1894). 'Beiträge zur Theorie der psychischen Analyse.' *Zeitschrift für Psychologie und Physiologie der Sinnesorgane* 6: 340–385.

Meskin, Aaron, and Weinberg, Jonathan M. (2003). 'Emotions, Fiction, and Cognitive Architecture.' *British Journal of Aesthetics* 43 (1): 18–34.

Metz, Christian (1972). *Semiologie des Films (Essais sur la Signification au Cinéma)*. München: Fink.

Montague, Michelle (2009). 'The Logic, Intentionality, and Phenomenology of Emotion.' *Philosophical Studies* 145 (2): 171–192.

Moran, Richard (1994). 'The Expression of Feeling in Imagination.' *Philosophical Review* 103 (1): 75–106.

Moyal-Sharrock, Danièle (2009). 'The Fiction of Paradox: Really Feeling for Anna Karenina.' In Y. Gustafsson, C. Kronqvist and M. McEachrane (eds.), *Emotions and Understanding: Wittgensteinian Perspectives.* Hampshire: Palgrave-Macmillan, 165–184.

Neill, Alex (1993). 'Fiction and the Emotions.' *American Philosophical Quarterly* 30 (1): 1–13.

Neill, Alex (2009). 'Art and Emotion.' In J. Levinson (ed.), *The Oxford Handbook of Aesthetics.* Oxford: Oxford University Press, 421–435.

Nichols, Shaun (2007). *The Architecture of the Imagination. New Essays on Pretence, Possibility, and Fiction.* Oxford: Oxford University Press.

Nussbaum, Martha C. (2001). *Upheavals of Thought: The Intelligence of Emotions.* Cambridge: Cambridge University Press.

Peacocke (1992). 'Scenarios, Concepts, and Perception.' In Crane, Tim (ed.). *The Contents of Experience.* New York: Cambridge University Press, 105–135.

Prinz, Jesse J. (2003). 'Emotions Embodied.' In R. Solomon (ed.), *Thinking about Feeling.* Oxford: Oxford University Press,1-14.

Prinz, Jesse J. (2004). *Gut Reactions: A Perceptual Theory of the Emotions.* Oxford: Oxford University Press.

Quine, Willard V. O. (1960). *Word & Object.* Cambridge: MIT.

Radford, Colin (1975). 'How Can We Be Moved by the Fate of Anna Karenina?' *Proceedings of the Aristotelian Society*, Supplementary Volume 49: 67–80.

Ratcliffe, Matthew (2005). 'The Feeling of Being.' *Journal of Consciousness Studies* 12 (8–10): 43–60.

Ratcliffe, Matthew (2008). *Feelings of Being: Phenomenology, Psychiatry and the Sense of Reality.* New York: Oxford University Press.

Robinson, Jenefer (2009). 'Intentionality (and Emotion).' In D. Sander and K. Scherer (eds.), *Oxford Companion to Emotion and the Affective Sciences.* Oxford: Oxford University Press, 219–220.

Ross, W. David (ed.) (1928). *The Works of Aristotle.* Vol. 1. Translated by E. M. Edghill. Oxford: Clarendon Press.

Schaper, Eva (1978). 'Fiction and the Suspension of Disbelief.' *British Journal of Aesthetics* 18 (1): 31–44.

Scholz, Oliver R. (2004). *Bild, Darstellung, Zeichen: Philosophische Theorien bildlicher Darstellung.* Vittorio Klostermann.

Searle, John R. (1975). 'A Taxonomy of Illocutionary Acts.' In K. Gunderson (ed.), *Language, Mind and Knowledge.* University of Minnesota Press, 344–369.

Searle, John R. (1977). 'Reiterating the Differences – A Reply to Derrida.' *Glyph* 1: 198–208.

Searle, John R. (1979a). *Expression and Meaning: Studies in the Theory of Speech Acts.* Cambridge: Cambridge University Press.

Searle, John R. (1979b). 'What Is an Intentional State?' *Mind*, LXXXVIII (349): 74–92.

Searle, John R. (1980a). 'Las Meninas and the Paradoxes of Pictorial Representation.' *Critical Inquiry* 6 (3): 477–488.

Searle, John R. (1980b). 'Intrinsic Intentionality.' *Behavior and Brain Sciences* 3: 450–457.

Searle, John R. (1980c). 'Analytic Philosophy and Mental Phenomena.' *Midwest Studies in Philosophy* 5: 405- 423.

Searle, John R. (1982). 'The Myth of the Computer.' *New York Review of Books* 29 (7): 3–6.

Searle, John R. (1983). *Intentionality: An Essay in the Philosophy of Mind.* Cambridge: Cambridge University Press.

Searle, John R. (1984). 'Intentionality and Its Place in Nature.' *Dialectica* 38 (2–3): 87–99.

Searle, John R. (1986). 'Meaning, Communication, and Representation.' In Grady and Warner (eds.) *Philosophical Grounds of Rationality: Intentions, Categories, Ends.* New York: Oxford University Press, 209–226.

Searle, John R. (1992). *The Rediscovery of the Mind.* Cambridge, MA: MIT Press.

Searle, John R. (1994). 'Structure and Intention in Language: A Reply to Knapp and Michaels.' *New Literary History* 25 (3): 677–681.

Shelley, Mary (1818/2018). *Frankenstein: The 1818 Text.* New York: Penguin.

Simons, Peter (1996). 'Logic in the Brentano School.' In L. Albertazzi, M. Libardi and R. Poli (eds.), *The School of Franz Brentano.* Kluwer Academic.

Slaby, Jan (2008a). 'Affective Intentionality and the Feeling Body.' *Phenomenology and the Cognitive Sciences* 7 (4): 429–444.

Slaby, Jan (2008b). *Gefühl und Weltbezug. Die menschliche Affektivität im Kontext einer neo-existenzialistischen Konzeption von Personalität.* Paderborn: Mentis.

Slaby, Jan (2012). 'Emotional Rationality and Feelings of Being.' In J. Fingerhut and S. Marienberg (eds.), *Feelings of Being Alive.* Berlin: De Gruyter, 55–78.

Slaby, Jan, Achim Stephan, Henrik Walter, and Sven Walter (eds.) (2011). *Affektive Intentionalität: Beiträge zur welterschließenden Funktion der menschlichen Gefühle.* Paderborn: Mentis.

Smith, Barry (1994). *Austrian Philosophy: The Legacy of Franz Brentano.* Chicago: Open Court.

Stecker, Robert (2010). *Aesthetics and the Philosophy of Art: An Introduction.* 2nd ed. Maryland: Rowman and Littlefield.

Stecker, Robert (2011). 'Should We Still Care about the Paradox of Fiction?' *British Journal of Aesthetics* 51 (3): 295–308.

Stocker, Michael (1983). 'Psychic Feelings: Their Importance and Irreducibility.' *Australian Journal of Philosophy* 61: 5–26.

Suits, David (2006). 'Really Believing in Fiction.' *Pacific Philosophical Quarterly* 87 (3): 369–386.

Taieb, Hamid (2017). 'Intentionality and Reference: A Brentanian Distinction.' *The Monist* 100 (1): 120–132.

Taieb, Hamid (2018). *Relational Intentionality: Brentano and the Aristotelian Tradition.* Springer International.

Takahashi K. and Watanabe K. (2015). 'Seeing Objects as Faces Enhances Object Detection.' *i-Perception* 6: 1–14.

Talmage, Catherine J. L. (1996). 'Meaning Intentions.' *Australasian Journal of Philosophy* 74 (2): 341–346.

Textor, Mark (2006). 'Brentano (and Some Neo-Brentanians) on Inner Consciousness.' *Dialectica* 60 (4): 411–432.

Textor, Mark (2013). 'Brentano on the Dual Relation of the Mental.' *Phenomenology and the Cognitive Sciences* 12 (3): 465–483.

Textor, Mark (2017). *Brentano's Mind.* Oxford: Oxford University Press.

Textor, Mark (2023). 'Brentano on Act, Content and Intentionality.' *Grazer Philosophische Studien* 100 (1–2): 173–196.

Van de Mosselaer, Nele (2018). 'How Can We Be Moved to Shoot Zombies?' *Journal of Literary Theory* 12 (2): 279–299.

Vendrell Ferran, Íngrid (2014). 'Das Paradoxon der Fiktion.' In T. Klauk and T. Köppe (eds.), *Fiktionalität. Ein interdisziplinäres Handbuch.* Berlin: De Gruyter, 313–337.

Vendrell Ferran, Íngrid (2016). 'Affective Intentionality: Early Phenomenological Contributions to a New Phenomenological Sociology.' In D. Moran and Th. Szanto (eds.), *Phenomenology of Sociality: Discovering the "We."* Routledge, 219–233.

Voss. J. L., et al. (2012). 'The Potato Chip Really Does Look Like Elvis! Neural Hallmarks of Conceptual Processing Associated with Finding Novel Shapes Subjectively Meaningful.' *Cerebral Cortex* 22: 2354–2364.

Wimsatt, William K., and Monroe C. Beardsley. (1946). 'The Intentional Fallacy.' *Sewanee Review* 54 (3): 468–488.

Walton, Kendall L. (1978). 'Fearing Fictions.' *Journal of Philosophy* 75 (1): 5–27.

Walton, Kendall L. (1990). *Mimesis as Make-Believe: On the Foundations of the Representational Arts.* Cambridge, MA: Harvard University Press.

Walton, Kendall L. (1997). 'Spelunking, Simulation and Slime.' In M. Hjort and S. Laver (eds.), *Emotion and the Arts*. Oxford: Oxford University Press, 37–49.

Werner, Christiana (2020), 'Emotions, Actions and Inclinations to Act.' *Erkenntnis*: 1–18.

Weston, Michael (1975). 'How Can We Be Moved by the Fate of Anna Karenina?' *Proceedings of the Aristotelian Society*, Supplementary Volume 49: 81–93.

Yanal, Robert (1994). 'The Paradox of Emotion and Fiction.' *Pacific Philosophical Quarterly* 75 (1): 54–75.

Yanal, Robert (1999). *Paradoxes of Emotion and Fiction*. University Park: Pennsylvania State University Press.

Zipfel, Frank (2012). 'Emotion und Fiktion. Zur Relevanz des Fiktions-Paradoxes für eine Theorie der Emotionalisierung in Literatur und Film.' In S. Poppe (ed.), *Emotionen in Literatur und Film*. Würzburg: Königshausen and Neumann, 127–153.

Zipfel, Frank (2018). 'Emotion, Darstellung, Fiktion. Literaturtheoretische Überlegungen zum Verhältnis zwischen Fiktionsparadox und Mimesisparadox.' *Journal of Literary Theory* 12 (2): 321–344.